University of Lowell
Alumni/Lydon Library

AF722798

handbook of electronic testing, measurement and troubleshooting

handbook of electronic testing, measurement and troubleshooting

MATTHEW MANDL

Professional Writer

Former Faculty Member
Technical Institute and Community College
Temple University

182307

Reston Publishing Company, Inc., Reston, Virginia
A Prentice-Hall Company

Library of Congress Cataloging in Publication Data

Mandl, Matthew.

Handbook of electronic testing, measurement, and troubleshooting.

Includes index.

1. Electronic apparatus and appliances—Maintenance and repair. 2. Electronic apparatus and appliances—Testing. 3. Electronic measurements. I. Title.

TK7870.M322 1976 621.381′028 75-16232

ISBN 0-87909-330-7

TK 7870 M322 1976

© 1976 by
Reston Publishing Company, Inc.
A Prentice-Hall Company
Reston, Virginia 22090

All rights reserved. No part of this book may be reproduced in any way, or by any means, without permission in writing from the publisher.

10 9 8 7 6 5 4 3 2 1

Printed in the United States of America.

contents

preface

This *Handbook of Electronic Testing, Measurement and Troubleshooting* describes the operational principles and applications of various testing and measuring devices in relation to the servicing and maintenance of the basic circuitry of electronics. Instruments under discussion include the voltmeter, ammeter, ohmmeter, oscilloscope and signal generators, as well as specialized equipment and typical commercial units. Instrument applications are not confined to a specific electronic system, but range from simple audio amplifiers to RF circuitry including solid-state units used in television. High-fidelity systems are also covered, including stereo, quadraphonics and associated gear.

Throughout this text the logical approach to electronic testing, measurement and trouble localization is stressed. Lengthy explanations of circuit theory have been avoided, though some basic descriptions of circuit function are provided where essential for proper instrumentation usage. Generally it is assumed the reader has a basic knowledge of electronics, including some dc and ac theory.

It is also assumed the reader has a fundamental knowledge of the bias factors relating to solid-state and tube electronics. For those requiring a review of these topics, however, the summary and related illustrations in Chapter 8 should be studied. Essential reference material such as resistor and capacitor color codes and other related data are included in the Appendix.

All line drawings and photographs are original with the author and represent the practical aspects of electronic testing, troubleshooting and maintenance.

To illustrate the practical aspects of electronic troubleshooting and use of testing principles covered in the text, Chapter 12 is devoted exclusively to more than three dozen case histories involving trouble localization of high-fidelity units (such as AM-FM receivers, stereo amplifiers, tape decks, etc.) and television receivers. Section 12-2 of Chapter 12 contains an Index for the topics covered in that chapter.

MATTHEW MANDL

1

fundamental factors

1-1 CLOSED AND OPEN CIRCUITS

A *closed* electric or electronic circuit is one wherein components are interconnected to provide for a closed circuit loop to permit current to flow. An *open circuit* is one in which the closed loop has been opened either by a switch arrangement, by a broken wire or component, or by the removal of a section of wire or a component. Thus, an open circuit can be deliberate or accidental.

Typical examples of open circuits are illustrated in Fig. 1-1. A battery, switch and light bulb (or doorbell, buzzer, etc.) are shown in Fig. 1-1A. With the switch open no current flows and the light is out. When the switch is closed a continuous circuit loop is formed and electrons flow from the negative terminal through the switch and light and to the positive battery (or generator) terminal. During such flow the current through the bulb generates heat and light. If the bulb burns out an undesired open circuit is formed. Similarly, a broken wire would form an undesired open circuit.

Other open-circuit conditions are illustrated in Fig. 1-1B. Here, assume that an audio amplifier has been turned on but no sound is pro-

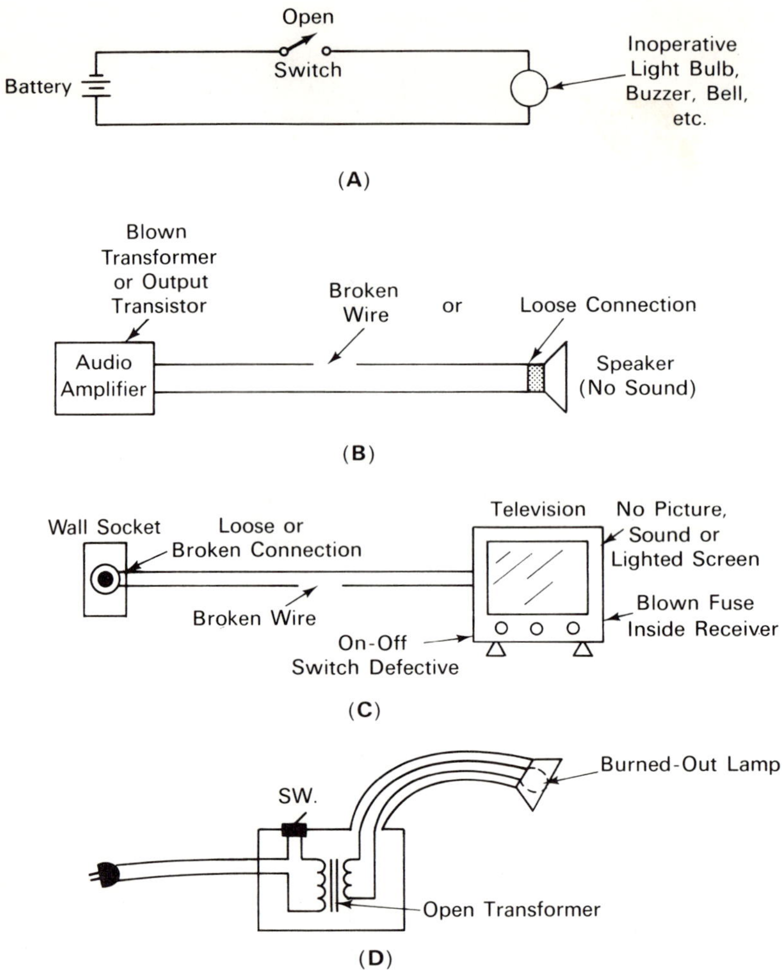

Figure 1-1 Examples of open circuits.

duced. If the pilot light is lit (or the dial light in case of a receiver) it is obvious that power is present. A dead speaker, however, could mean an open circuit formed by a broken wire, a loose connection at the speaker terminals, or a defective output transformer or output transistor. It is also possible that the speaker was overloaded by excessive audio power and the voice-coil windings burned out. If a visual inspection fails to disclose the open-circuit condition, a substitute speaker (if readily available) can be plugged into the amplifier. This will indicate whether the trouble is in the speaker wire and speaker system or the amplifier. Thus, if connecting another speaker fails to produce sound, the audio amplifier must be checked (see Chapters 3 and 6. Also refer to the Index for cross-reference material).

Figure 1-1C shows an inoperative television receiver. Presence of *any* type sound from the speaker (music, noise, hum, etc.) would indicate that power is arriving at the set. Also, any light produced internally in the picture tube and appearing on the screen indicates power arriving at the set. Absence of sound and picture might indicate a blown fuse within the receiver, a broken wire in the ac line cord, or a loose or open connection in the wall socket. The on-off switch of the receiver could also be defective. By plugging a lamp (known to be operating properly) into the socket, the presence of socket power can be ascertained.

If the lamp fails to light the problem lies with the ac socket. (One must not overlook the possibility, even though remote, that both the socket and the ac cord could be defective.) If the lamp does light, the trouble will be found in the parts associated with the receiver such as the ac plug, cord, internal fuses, on-off switches and so forth. These items can be checked out as detailed in subsequent chapters, including voltage checks, ohmmeter readings and continuity tests.

A final example, shown in Fig. 1-1D, is that of a high-intensity desk lamp. Here a small transformer in the lamp base steps down the 120 V ac to 6 V or 12 V. Open circuits can occur in the plug, line cord, internal transformer and light bulb.

1-2 SHORT CIRCUITS

The *short circuit* (or the partial shorts described in Sec. 1-3) is an undesired condition in an electric or electronic circuit that can cause damage and use excessive electric power. A short (or partial short) occurs when the circuit resistance is abnormally low in ohmic value compared to the power source used. If, for instance, a copper bar is placed directly across the terminals of a battery, the virtually zero resistance of the copper strip permits such a high current conductivity that all the current flows that the battery can furnish. The bar of copper, in such an instance, "shorts" out the battery and causes the so-called *short circuit.* Short circuits cause damage to the battery, generator or other voltage source, hence protective devices are used to open the circuits when excessive currents flow. Thus, when a short circuit occurs it causes fuses to blow, or opens circuit breakers, and thus prevents a severe and possibly damaging load on the power source.

Whenever a fuse has opened or circuit breaker has tripped, the cause should be ascertained before the fuse is replaced or the circuit breaker reset. On occasion a voltage surge may trip a circuit breaker or open a fuse (particularly if it is not of the "slow-blow" type) and no short circuit exists. If, however, a short circuit does exist, fuse replacement or circuit breaker resetting will result only in another blown fuse or a re-opening of the circuit breaker.

1-3 LEAKS AND PARTIAL SHORTS

When a resistance is placed across a battery the amount of current flow depends on the battery voltage and the ohmic value of the resistance, since I (current) = E (voltage) divided by R (resistance). As the resistance value is lowered, more and more current flows and the load on the battery increases. If the resistance continues to decrease a point is reached at which an excessive amount of current flows from the battery, the battery overheats and damage occurs. At the point where overheating occurs the battery is considered to be *partially shorted.*

Components such as tubes, transistors, capacitors, transformers, coils and resistors, can develop defects and cause leakage currents to flow or create partial shorts in the circuits to which they are connected. A typical example is shown in Fig. 1-2A where two noise-filter capacitors are shown across the primary of a transformer used to furnish power to such devices as radios, high-fidelity units and television receivers. Since such capacitors are used across the ac line they are of the nonpolarized type with ratings ranging from 0.0001 to 0.001 μF, depending on whether RF or audio noise is to be filtered. Such capacitors have a voltage rating above the peak of the line voltage ($1.41 \times \text{rms} = 1.41 \times 120 = 168$ V) but improper humidity, shielding, defects during manufacture or deterioration because of age can cause leakage.

Until such capacitors become shorted or partially shorted, slight leakage wll affect filtering only. In time, however, the leakage grows and an increasing drain on ac power occurs. A capacitor checker or an ohmmeter test for leakage will disclose the defect. A shorted capacitor will affect fuses or circuit breakers.

Shorts can also occur in wall sockets, in wiring because of poor or broken insulation, and in switches as shown in Fig. 1-2B. Again, such shorts or partial shorts will cause fuse failure or circuit-breaker tripping and, as mentioned earlier, fuses should not be replaced or circuit breakers reset until the short has been removed.

Not all shorts damage equipment or cause fuses to blow. A typical example is shown in Fig. 1-2C, where a short is shown in the wires of the phonograph tone arm. Here the short would prevent the transfer of sound to the amplifier, but would not cause excessive currents because of the low signal voltage produced by the phonograph cartridge.

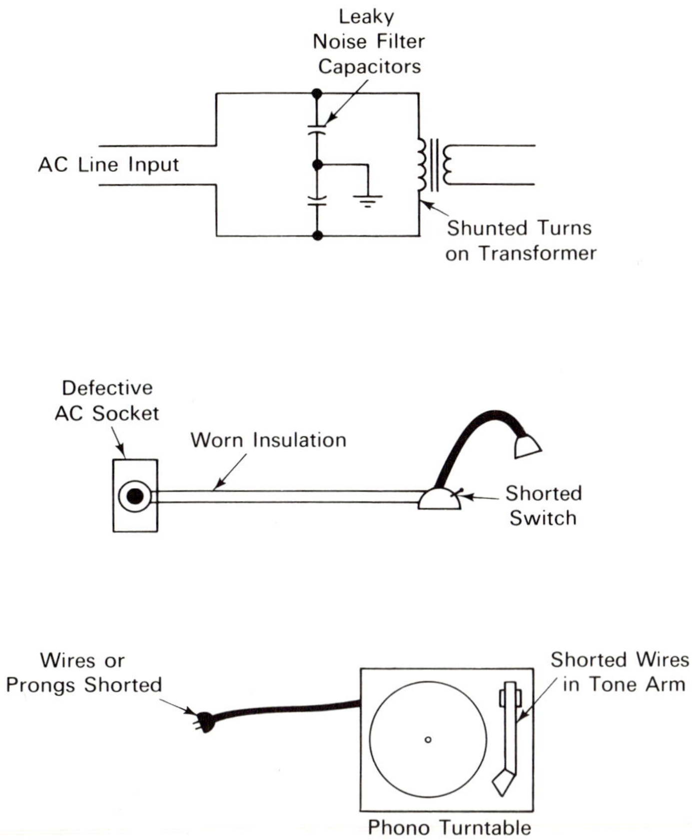

Figure 1-2 Shorts, leaks, and partial shorts.

1-4 INITIAL TESTS AND CHECKS

The major contributing causes of poor performance in electronic equipment are open, shorted or partially shorted components that weaken or distort the signals being handled. Thus, defective components alter currents and voltages by changing the ohmic or reactive values of circuits. In many instances, however, poor performance in electronic or electric units is caused by misadjusted controls rather than component failure. A representative example is that shown in Fig. 1-3 where the loss of horizontal sync has occurred in a television receiver. Most of us know by now that this can usually be corrected by a simple adjustment of the horizontal hold control. If this fails to correct the trouble the coarse frequency control in the horizontal oscillator section must be adjusted to bring the hold control within the adjustable range (see also Chapter 10).

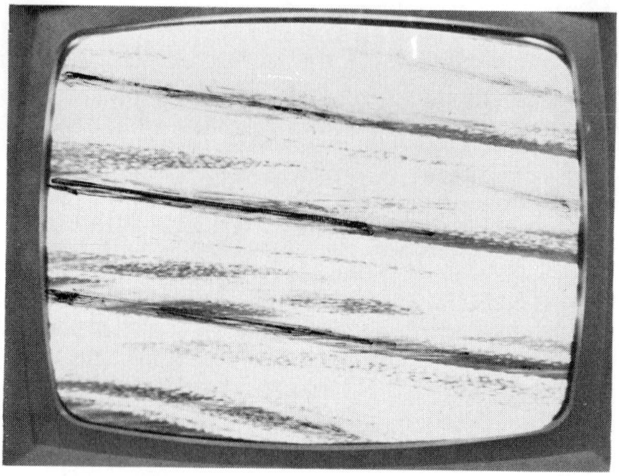

Figure 1-3 Loss of horizontal sync in TV.

In a television receiver (particularly color) there are many more controls (both front panel and rear or hidden) that affect picture quality including tint, color, automatic gain control and color killer. Similarly, high-fidelity stereo or quadraphonic equipment have bass and treble controls and balance controls. In addition speakers must be properly phased and certain maintenance procedures must be undertaken periodically to keep the units in peak operating condition (see Section 1-5 and 1-6).

Thus, for poorly operating equipment, the experienced technician makes an original diagnosis by visual inspection and by listening to sound quality. On occasion he may notice certain odors that indicate overheating of transformers, wires or other components because of partial or complete short circuits. Logic often dictates initial test procedures that can be performed without specific test equipment.

As an illustration, assume a radio is inoperative, and has features such as shown in Fig. 1-4A. If this receiver is turned on for battery operation, but the rear switch is on ac the switch should be thrown to battery operation. If the sound is still not forthcoming the battery compartment should be checked for proper battery placement, for tight-fitting holders and for weak batteries.

Another possibility for no sound is a defective earphone socket. In many instances the earphone socket is of the type that disconnects the speaker when the earphones are plugged in, as shown in Fig. 1-4B. When the earphones are removed the switch contacts must close for speaker operation. For a quick check, earphones can be inserted and checked for sound output. If sound is heard in the earphones but not from the

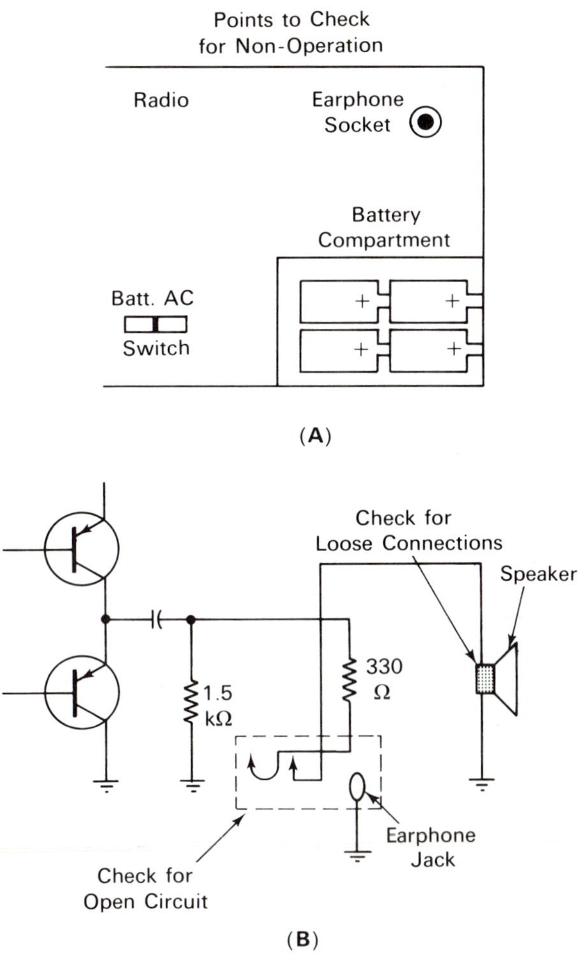

Figure 1-4 Potential trouble spots in radio.

speaker, check for a defective socket switch (or for possible loose connections to the speaker terminals).

For units operating on the ac line, an inoperative device should be tried in a different socket to make certain that the initial socket is not defective and thus not furnishing line voltage. (A quick check is to plug a lamp into the socket.) If the ac socket is functioning a visual inspection of the plug and cord is indicated. Next the on-off switch is checked for possible inoperation. If these initial checks fail to disclose the trouble then more extensive tests and measurements must be undertaken as detailed in subsequent chapters.

For weak operation of AM or FM receivers and television sets, atmospheric conditions may be contributing to temporary difficulties, or the general area may be one of weak signals. Thus, for improved reception an outside antenna may be necessary, or an existing one checked for possible open circuit in the lead-in or at the antenna elements. If only one station in many is weak it is unlikely that the fault is in the receiver or antenna, but may be a temporary condition at the transmitter. As a preliminary test, an indoor antenna may be tried and compared with the existing antenna system.

For television receivers the indoor antenna check is useful to determine if trouble exists with the outdoor antenna. Obviously if reception with the indoor antenna is as good (or better) than that obtained from the outside antenna, the latter is either defective or of improper design (or not oriented properly toward the stations to be received).

Another quick check is to disconnect the two wires of the twin lead from the antenna terminals of the receiver. Using only one wire of the twin lead touch first one antenna input terminal, then the other, while checking reception for the strongest station to be received. Repeat this procedure with the other wire of the twin lead. Results should be equal for both wires, even though general reception will not be as good as when both wires are connected. A dissimilarity between the two indicates an unbalance in the antenna system. Such an unbalance could be caused by one of the twin-lead lines having an open circuit along its length (or at its connection to the antenna elements). Also, some antenna elements may be missing (often the case with older antennas that have been exposed to high winds and corrosion over the years). Since antenna elements are usually symmetrically arranged along each side of the holding bar, a visual inspection discloses such faults.

Construction of high-rise buildings in the area can affect television reception, particularly from distant stations. In such instances the antenna may have to be raised higher, or a more powerful one installed to overcome weak-signal conditions. Since proper orientation is so important, a motorized rotator is of considerable help when all stations are not in the same direction.

Antennas come in a variety of sizes and designs to suit particular needs. For fringe areas, booster amplifiers for VHF and UHF are also available for improving reception. (Refer to the Index for additional data, and see also Chapter 10.)

1-5 TAPE MACHINE MAINTENANCE

All magnetic-tape machines, because of the many moving parts and the sliding of the tape across erase, record and play-back heads, require periodic maintenance. Thus, to restore peak operation some parts must be oiled on occasion and others cleaned. Motors must be oiled to minimize wear and noise and to maintain proper speed for the particular machine in use.

Open-reel tape recorders operate at speeds of either 3¼ or 7½ inches per second (ips), though some machines may have even slower speeds in specialized applications. Cartridge tapes are ¼ inch wide but the cartridge tapes run 3¼ inches per second. The cassette tape width is slightly more than ⅛ inch (0.146″) whereas that of the open-reel tape is slightly less than ¼ inch (0.246″). Cassette tape speed is 1⅞ inches per second. Both the cassette and cartridge tapes are enclosed in a housing containing its own spools. Cassette units can be wound fast in forward or reverse. The cartridge units are continuous-operation types with the tape forming a continuous loop within the housing.

Cassettes record and play a stereo section of dual tracks side by side as shown in Fig. 1-5A, thus permitting playback on a mono machine. Reel-to-reel tapes are not compatible because the stereo pairs are interlaced between the pair of the second track in four-track operation, as shown in Fig. 1-5B. Hence stereo-recorded open reel tape cannot be played back on a mono machine.

For proper tape operation and maximum pick-up sensitivity the coated side of the tape slides past the heads (usually under slight pressure

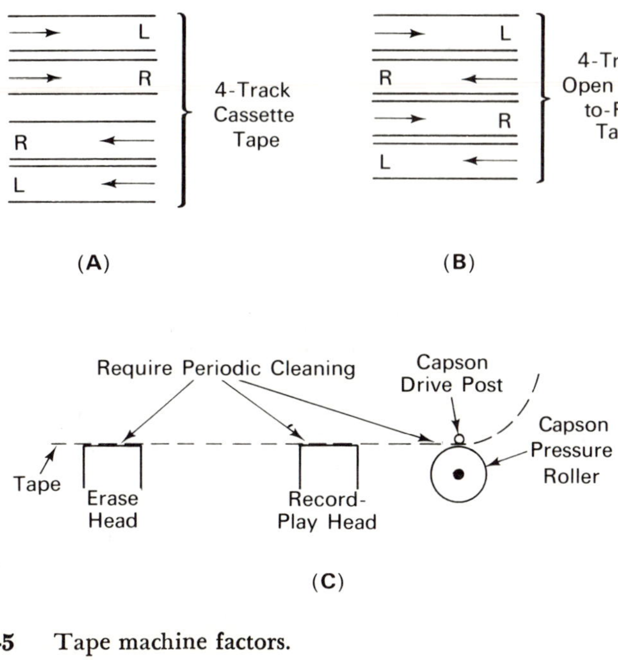

Figure 1-5 Tape machine factors.

from spring-loaded felt pads). Depending on the quality of the tape used, a certain amount of material rubs off on the heads, eventually clogging them because of the tape oxide material that collects in the gap of the playback and recording heads. The results are fuzzy tonal quality plus weak sound, with loss of high frequency tones. Because of the passage past the heads of magnetized areas of recorded sound, the heads sometimes develop residual magnetism and should be demagnetized by degaussing coils (available from jobbers).

Most of the common troubles with tape machines arise because of the need for head cleaning and lubrication of moving parts. When lubricating the capstan pressure-roller bearing and other parts, keep oil away from roller surfaces (and belts inside the tape machine). Cotton swabs dipped in rubbing alcohol (or cleaning fluid available from radio stores) are useful for cleaning heads in tape machines. Jobbers also sell long-stemmed cotton swabs for reaching into the compartments of tape cartridge units. During the head-cleaning process the capstan drive roller should also be cleaned (see Fig. 1-5C). Slippage at the capstan pinch roller results in uneven speed and wavering sound.

Some higher-priced units have automatic head demagnetization devices built in. Usually, however, ordinary units must be demagnetized on occasion, though this need not be done as often as head cleaning. If poor results are obtained from a tape machine, however, the head demagnetization should not be neglected.

1-6 PHONO MAINTENANCE FACTORS

When a phonograph record produces a wavering and uneven sound it is usually an indication that a thorough lubrication of the moving parts (including the motor) is required. A light machine oil is preferred to assure free-moving action of the tone arm in both the lateral and horizontal movement. If the instruction manual specifies the type oil to be used for specific parts the instructions should be followed. Keep all oil away from any rollers or rubber mounts.

If uneveness of reproduction is only slight or if the listener is uncertain of performance, a precise check can be made using a stroboscopic disk as shown in Fig. 1-6. Such a disk is placed on the phonograph turntable and the machine turned on. The disk is viewed under a neon or fluorescent light. The viewer selects the circle that corresponds to the speed used. Thus, if the machine is operating at 45 rpm the outer circle segments will appear to be standing still. If a speed control is present it can be adjusted for exact speed. If the turntable runs above 45 rpm the segments will appear to move, with the segment speed determined by

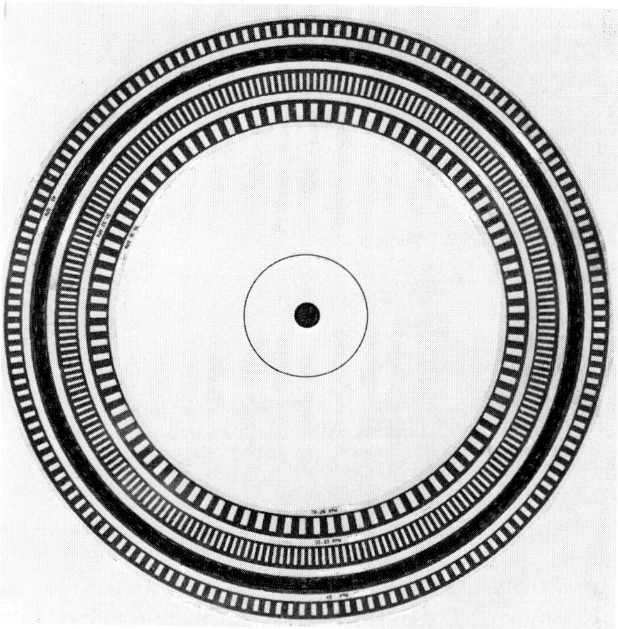

Figure 1-6 Stroboscopic disk.

how much faster the turntable rotates. For speeds below 45 rpm, the segments appear to move in the opposite direction.

Even with no speed control (and assuming the actual speed is slightly off) the movement of the segments should be steady and unfluctuating. If the segments move rapidly lubrication is required. It is also possible that the drive rollers are worn or slipping and these should be cleaned with alcohol and checked for wear.

The second test circle of the strobe disk is for 33.3 rpm records and this has the most segments. The inner circle is for checking the older machines having a 78 rpm speed. Some phonograph turntables have a strobe disk as a permanent part of the turntable section. With all such devices, however, poor results are obtained under an incandescent light; a fluorescent type provides maximum visibility of segment indications.

Lint and dust tend to accumulate in record grooves and eventually clog the needle tip. The result is a fuzziness to the sound with loss of high-frequency tones. Thus, such symptoms call for a check of the needle point and removal of all accumulated lint (see Fig. 1-7A). Use a small camel-hair brush and gentle forward-and-back strokes rather than brushing the needle from side to side. If trouble is experienced in ridding the needle point of accumulated lint, dip the brush in alcohol or lighter

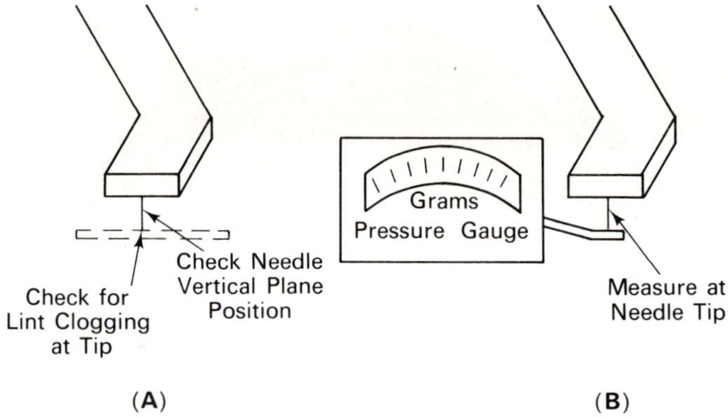

Figure 1-7 Phone cartridge checks.

fluid and try again. Special brushes and record-cleaning cloths are available for cleaning purposes.

Fuzzy high tones and poor quality, of course, may also be caused by worn records. Another cause is a worn needle that, in turn, will also damage record grooves. The needle point is difficult to evaluate by visual inspection. Small 50-power pocket-size magnifiers are available from most high-fidelity outlets and these provide a more accurate means for checking needle point. A replacement needle (if available) could be tried as a sure check for a worn or defective needle.

If needle and record wear are excessive, the needle pressure on the records should be reduced to the minimum adjustment without the needle jumping grooves during play. The term *tracking force* is used to designate the grams of needle pressure. If the tracking force of a phono cartridge is given as from 3 grams to 6 grams, the weight adjustment of the arm should be set for 3 or 3½ grams unless the particular design of the phono arm and turntable does not permit good tracking at this pressure. In this respect the level of the turntable should be checked because a slight tilt will increase record-groove jumping or cartridge skating across the record. Use a small carpenter's level at front and sides of the turntable and adjust for any tilt by resetting the bolt-spring nuts (if present) or by placing cardboard sections under the sides of the housing.

Good leveling of the turntable is of considerable importance when the cartridge tracking force is around one or two grams. If the turntable has an antiskate device make sure it is set to conform to the tracking force of the stylus in use. If the cartridge has a stylus brush attached to its front, the weight may have to be readjusted to compensate for the slight pressure exerted by the brush.

In the higher-priced models a tracking force around one gram is possible for minimum record wear and increased stylus life. The pressure should be checked with a pressure gauge at the needle tip (see Fig. 1-7B). Record wear increases appreciably at tracking forces above two grams.

Magnetic cartridges (as opposed to the crystal or ceramic types) requires a preamplifier (either a separate unit or built into the amplifier circuit). Thus, low output may be the result of a mismatch by using a low-sensitivity cartridge with a low-grain amplifier. When replacing cartridges make sure this factor is taken into consideration. Also check for designated tracking force, frequency response and compliance when replacing old cartridges or needles. The compliance is the ability of the needle to move freely from side to side, thus *complying* with the record groove variations. A cartridge rating of 35×10^{-6} has excellent compliance and one of 15×10^{-6} is good. A frequency response of at least 20 Hz to 20 kHz will provide high-quality music.

1-7 TRANSFORMER TROUBLES AND TESTS

When transformers overheat or fail to function properly the symptoms are often misleading. In many cases it is not the transformer that is at fault, but rather the associated components which may cause transformer overloading or malfunctioning. Thus, when an abnormally warm or hot transformer is found, tests should be made on associated capacitors, resistors, transistors, tubes and other components to see that they are not drawing excessive current from the power supply (through the transformer) and thus overloading the normal current-carrying capacity of the transformer.

A typical example is shown in Fig. 1-8A, which illustrates a transistorized audio output stage. Here the transformer primary (L_1) transfers the dc potential of the power supply to the collector of the transistor. The audio signals developed between the collector and emitter circulate in the primary and are coupled to the secondary winding (L_2) and to the voice-coil of the speaker. Capacitor C_2 filters high-frequency transients by providing a low reactance for the signals in the extremes of the high-frequency audio range. If this capacitor should short out it would provide a direct path to ground for the power-supply potential applied to the bottom of L_1. Consequently, excessive currents would circulate in the transformer primary and cause overheating.

If capacitor C_2 is leaky or shorted, one having the same ohmic value must be used as a replacement. One with a higher voltage rating than the original can be used, however, to minimize breakdown. Thus, if

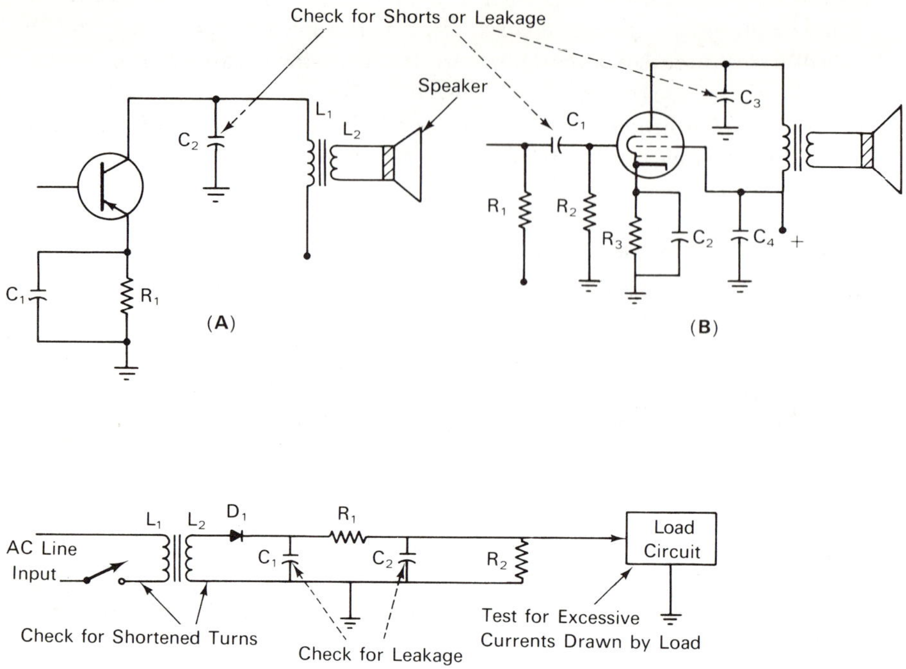

Figure 1-8 Troubles affecting transformers.

C_2 has a value of 0.002 Ω and is rated at 100 V, a 0.002 Ω, 150 V would be preferable to avoid future trouble. If the sound quality is poor after C_2 has been replaced it may indicate transformer damage because of the overheating. This can cause turns to become shorted, insulation to burn through, or an open circuit condition. Also, the power supply could have been damaged because of overload. Thus, additional checks will have to be made as detailed in subsequent chapters, including ohmmeter checks of the transformer, voltage checks on the power supply, and continuity checks for open circuits.

A tube version of the output audio amplifier is shown in Fig. 1-8B. If the coupling capacitor in the grid of the tube becomes leaky or shorted, a positive voltage will be applied to the grid and cause excessive currents to flow in the tube. Consequently the primary of the output transformer again heats up and becomes overloaded. Also, capacitor C_3, as with C_2, can cause difficulties. Changes in values of capacitors and resistors in the emitter circuit will upset bias potentials and thermal stability and cause excessive circulating currents. Thus, in Fig. 1-8A, capacitor C_1 and resistor R_1 should also be tested. In Fig. 1-8B, resistor R_3 and capacitor C_2 regulate tube bias, hence defects can cause excessive currents.

Capacitor C_4 in Fig. 1-8B is known as a *decoupling capacitor.* It filters hum for the screen grid and also isolates anode and grid by placing the screen grid at ground potential. (A capacitor has an increasingly lower shunt reactance for signals with higher frequencies. Thus, the screen is placed at ground for the *signal.*) If this capacitor opens, oscillations may occur because of coupling between output and input. Also, the hum level will increase. If C_4 shorts, it will overload the power supply though will not usually damage the transformer.

Overheated transformers give off a distinctive odor of burning insulation and the technician soon learns to recognize the symptom. Shorted turns will cause overheating because of the reduction of winding impedance. The transformer, however, unless originally of poor design, or slightly overloaded over a long period of time because of poor circuit design, will rarely cause trouble. Once overloaded, however, checks should be made for shorted turns, particularly in power supplies of the type shown in Fig. 1-8C. With low-voltage units shorted turns are difficult to check because of low ohmic values. Continuity is still present, hence an accurate low-reading ohmmeter must be used and readings checked against manufacturer's specifications if available. Otherwise, a new transformer will have to be installed if all other checks seem to indicate shorted turns. A simple check is to disconnect the load from across L_2 and apply ac to the primary. The transformer should remain cool (or only become slightly warm) even after a half hour or so of no-load operation. Overheating without a load connected is a sure sign of shorted turns in either the primary or secondary winding or both.

A common overloading that occurs in power supplies is the development of a leak or short in the filter capacitors. If capacitor C_1 develops a leakage it will overload both the rectifier diode D_1 as well as the transformer. (Both primary and secondary are overloaded, because any excessive currents drawn from L_2 must be obtained from the primary winding L_1.) A leaky or shorted C_2 will not apply as much of an overload as C_1 because of the series resistor R_1 which reduces the amount of current which would flow otherwise. Besides the two filter capacitors, overloads can also occur because of shorts or partial shorts within the load circuit (radio or audio amplified circuitry, etc.).

A fuse present in either the primary or secondary (or both) circuits, affords protection against damage by overload. Thus, if a fuse has opened, the circuitry should be checked for abnormally low resistance across the secondary winding L_2. If replacement fuses burn out it is a sure sign that some overload condition is present. The exception would be the incorrect use of a standard fuse instead of the slow-blow type.

Slow-blow fuses are used to prevent momentary transient voltage peaks or momentary noise pulses from blowing the instant-opening fuse.

A shorted rectifier D_1 (in which a low resistance is present in both the forward and reverse bias directions) will overload the filter capacitors because these are polarized and will overheat when ac is applied instead of dc. Thus, if the filter capacitors are defective a check should be made of the rectifier diode(s) (see also Sec. 2.9).

1-8 GROUNDS AND GROUNDING

The term *ground* generally applies to a common connection in receivers, amplifiers, tape players and other electronic gear. For most of these items the ground is the metal chassis. In electrical wiring systems the word *ground* is literal because the earth itself is used via water pipes or metal rods driven into ground. As shown in Fig. 1-9A, an ac cable in home wiring (117 V) will have three wires in it, even though the particular wiring may be a so-called two-wire system. The three wires are housed in a protective sheath of insulating material, and two of the inner wires are also insulated. The third wire (usually bare) is for grounding purposes.

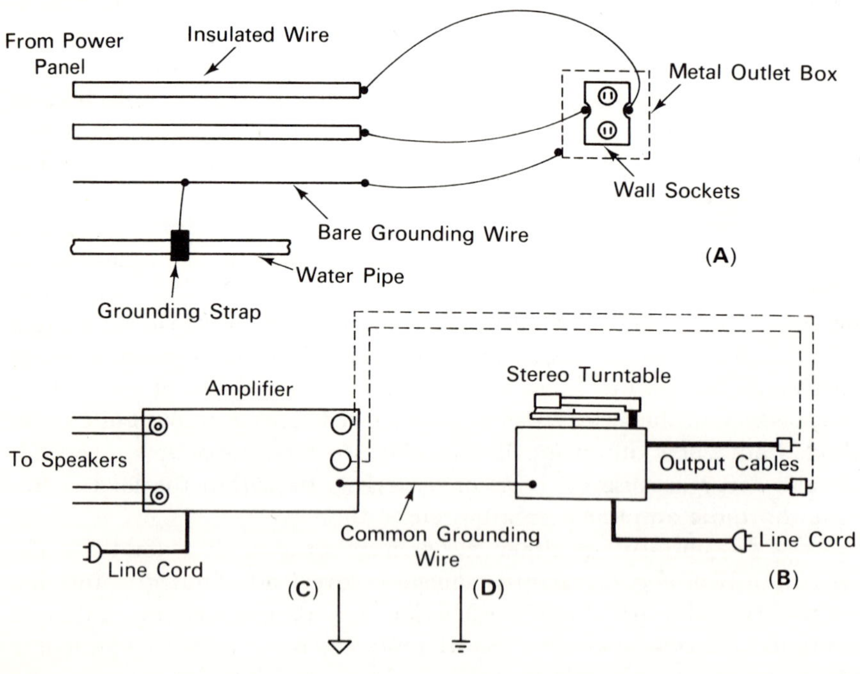

Figure 1-9 Ground connections.

When this ground wire is attached to the outlet box as shown it provides protection should one of the so-called *hot* wires short to the case. The ground wire would then short out the hot line and either blow a fuse or trip a circuit breaker. In most instances one of the insulated wires would also be placed at ground (water-pipe connection), but the additional ground wire is a standard safety precaution.

Ground connections between electronic units are important to establish a common return path for current flow. For most phonograph turntables, for instance, a single wire is provided for grounding the turntable mechanism to the chassis of the amplifier or stereo radio amplifier sections. Without such a grounding wire a high hum level may be present when magnetic pickups are used. (Some hum will also appear with the more sensitive crystal or ceramic cartridges.) Thus, when hum is present such an interconnection may be missing or the wire may have come loose. A check should be made to see that a good-contact ground linkage exists.

It would appear that the ground wire is not necessary because the outer conductors of the coaxial-cable audio connectors establish a common ground. While this is true for any dc component, the audio signals (and the 60 Hz line) are alternating in character and when fairly lengthy cables are used (over a foot or so) ground loops are set up because of inductive reactances established in the cable wires. In addition, there is a capacitive reactance established between the inner conductor of the coaxial cable and its outer conductor, again having an influence on ac signals. Thus, a separate ground wire is essential in many installations of this type.

The symbol shown in Fig. 1-9C refers to a ground connection that is *above* the common chassis connection. This symbol is used when the common connections of a circuit are isolated from chassis by a capacitance to prevent possible short circuits. The short could occur if the chassis has a high potential on it in relation to common ground connections. This condition was common with the low-priced tube radios in which the 117 V ac line was applied to the power supply without a transformer. Thus, one side of the ac connected to the chassis and, depending on which way the plug was inserted into the wall socket, the chassis could be *hot* with respect to ground. Where the symbol shown in Fig. 1-9D is used, it should indicate a common chassis ground.

Care also must be taken when using test instruments to avoid incorrect grounding of circuitry. If, for instance, three resistors are across the output of a power supply as shown in Fig. 1-10 (with R_3 connecting to chassis ground as shown, precautions must be observed for some test equipment. If tests are to be made across R_2 alone with a voltmeter a proper reading would be obtained. If, however, an oscilloscope is used and the lower test probe is at ground, a common ground linkage between

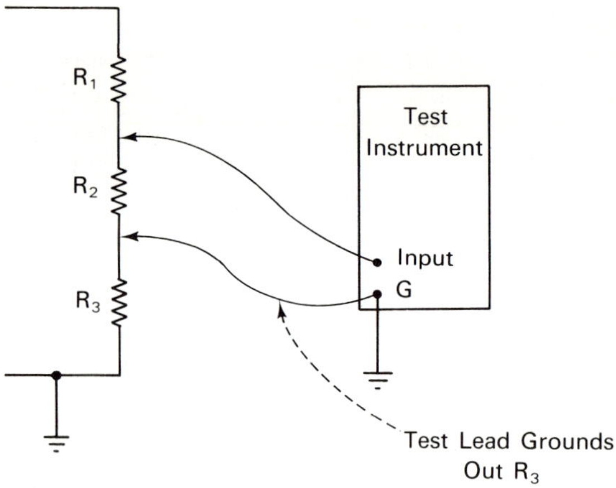

Figure 1-10 Incorrect test procedure.

the power supply and the scope would short out resistor R_3. Similar precautions must be taken with other equipment such as signal generators and dot generators in which one of the input lines is a ground potential (such as the outer shield of a coaxial-line input or output cable).

2

checking current and resistance

2-1 D'ARSONVAL METER UNIT

Most general-purpose meter movements use the *D'Arsonval* principle (named after the inventor who worked on it in 1880). The basic unit is illustrated in Fig. 2-1 and consists of a permanent magnet with two soft-iron pole pieces to concentrate the magnetic flux in the area where a moving coil is pivoted. The coil is wound on an aluminum form whose shafts usually pivot in jeweled (sapphire) bearings. Flat coil springs are mounted at each end and are wound in opposite directions to provide a countertorque against the coil-form movement when the meter is in use. The springs also act as lead-in wires for channeling the electric current through the coil. With current flow through the coil an electromagnet is formed with poles either aiding or opposing the fields of the permanent magnet. Thus, the necessary torque is developed to cause pointer movement to either left or right.

The aluminum frame upon which the coil is wound also acts as a damper system to prevent rapid oscillations of the needle before it comes to rest. Current is induced into the aluminum frame and an electromotive force is generated which opposes that in existence, thus providing effective damping of the pointer.

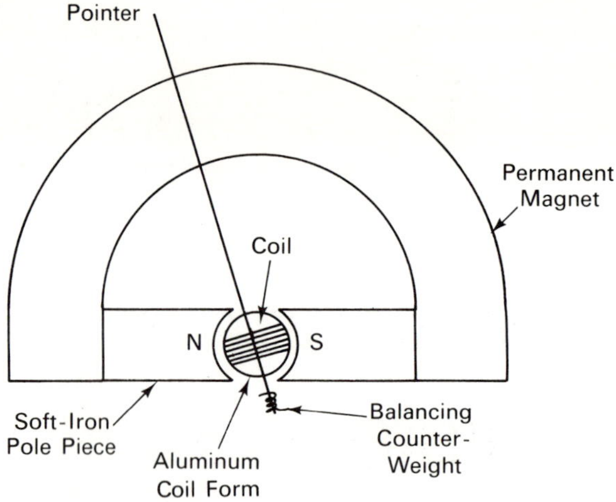

Figure 2-1 Basic D'Arsonval meter movement.

2-2 BASIC AMMETER

The basic meter described in Sec. 2-1 is essentially a current-reading device. Thus, the dial face of such a meter would be calibrated in amperes to read units of current, or milliamperes to read fractional values. Other instruments such as voltmeters and ohmmeters use the basic current meter with external circuitry as discussed later in this chapter.

The ammeter has a low internal resistance, the value of which depends on design factors such as the degree of accuracy needed and the ranges required. As opposed to the voltmeter (which is placed across a voltage to be measured) the ammeter is inserted in series with the circuit. In some special instances current measurements can be made without opening the circuit by employing induction principles as described later.

For the basic current-reading meter, the needle deflection is proportional to the amount of current flow through the meter coil (up to the maximum deflection of the meter scale). Thus, if maximum pointer deflection to the right indicates 1 A of current, then the needle would point to fractional current values as it moved to the left of the maximum value.

The range of values to be read can be increased by shunting resistors across the meter movement. Each shunting resistor bypasses a required amount of current around the meter. The meter scale is changed so that when a particular shunt resistor is used the appropriate scale can be read to indicate the new meter range. Commercial meters incorporate a rotary switch that has several positions for selection of the required shunt resistor and current-reading range.

The basic switching circuit shown in Fig. 2-2 has four shunt resistors. More could be used to increase the meter range into the ammeter value readings as well as milliamperes. The switch shown in Fig. 2-2 has five positions. The first position consists of an open circuit without shunt resistance. Thus, if this particular meter is a 1 mA basic movement and no shunt is used, the meter would read fractions of a milliampere up to the full-scale deflection of 1 mA. With shunt resistors thrown into the circuit, full-scale readings could be 0 to 5 mA, or 0 to 10 mA and so forth. If a resistor provides a 0 to 10 mA scale, only 1 mA would still flow through the meter coil, but 9 mA would be shunted by the selected resistor.

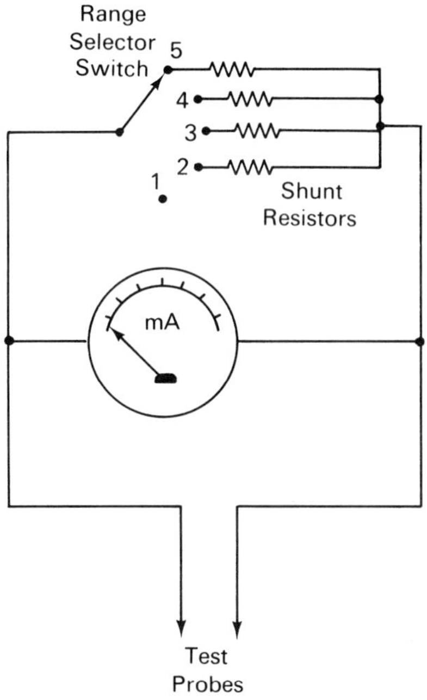

Figure 2-2 Basic mA circuit.

For meters having other basic movements, changes in the value of the shunt resistors would be necessary. The basic meter coil has a current limit for full-scale deflection and for a higher scale reading some current must be shunted around the meter. If, for instance, we wish to extend the range of a 0 to 3 mA meter to read a full-scale of 10 mA, the shunt resistor would have to pass 7 mA so that the 3 mA maximum is not exceeded through the meter coil. (Excessive currents would cause the pointer to swing beyond full scale with possible damage, and the meter coil could overheat and burn out.)

If the meter coil has a resistance of 14 Ω, the voltage drop across the meter coil at full-scale deflection (3 mA in our example) would be

$$E_c = IR = 0.003 \times 14 = 0.042 \text{ V}$$

Ohm's law can now be used to find the value of the required shunt resistor because the latter is in parallel with the meter coil and has the same voltage across it:

$$R_{sh} = E/I_{sh} = 0.042/0.007 = 6\ \Omega$$

It is convenient to express the resistance relationship and the amount by which the scale is to be multiplied in an equation. The value of the required shunt resistor can thus be found for any full-scale reading required, provided the ohmic value of the internal meter resistance is known. The equation for finding the value of the shunt resistor for ammeters or milliammeters is:

$$R_{sh} = R_m/(N - 1) \tag{2-1}$$

where R_{sh} is the ohmic value of the shunt resistor
R_m is the internal resistance of the meter
N is the number by which the meter scale is then to be increased

(For Equation 2-1 it must be noted that N indicates *the number by which the scale is to be increased—not* the full-scale range.)

Example: A milliameter with a basic 0 to 1 mA movement has an internal resistance of 45 Ω. The scale is to be increased to read a maximum current of 10 mA. What is the ohmic value of the required shunt resistor?

Answer: $45/(10 - 1) = 45/9 = 5\ \Omega$

2-3 THE GALVANOMETER

The galvanometer, as with the milliammeter, is a basic fractional current reading meter. In general usage, however, the galvanometer is not employed to show the actual unit values of current flowing through it, but instead indicates a proportional positive or negative *change* in a circuit. It is primarily a sensitive laboratory instrument usually of such a design that the zero of the scale is at the center, as shown in Fig. 2-3. The negative changes in a current direction are to the left of the center zero, and the positive changes to the right.

The galvanometer is often used in bridge circuits as discussed in Sec. 5-2. Thus, it acts as a sensitive device indicating polarity change, or as a meter for showing deviations from zero center without regard for particular current values.

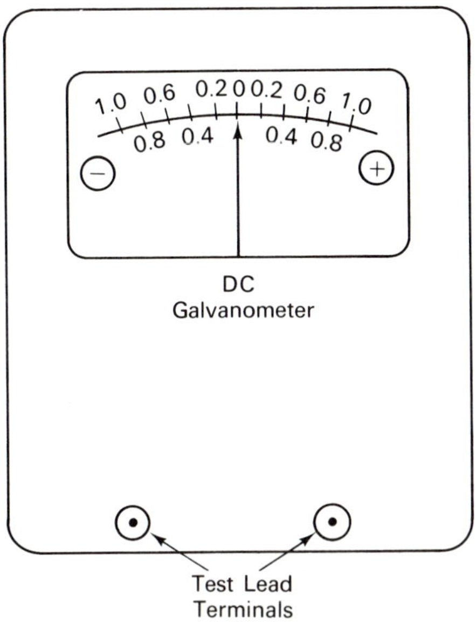

Figure 2-3 Zero-center galvanometer.

2-4 RECTIFIER-TYPE (AC) METERS

The basic dc meters can be converted for usage on ac by employing a diode rectifier as shown in Fig. 2-4. Calibration, sensitivity and current-carrying capacities become problems for reading alternating *currents,* hence the basic rectifier-type meter is primarily used in *voltmeter* applications. (See also Sec. 3-3.) Essentially, however, the basic ammeter principle still applies, though the values resulting from rectification must be understood when using or repairing such meters.

Typical meter rectification circuitry is shown in Fig. 2-4. In the half-wave system, seen in Fig. 2-4A, diode D_1 permits current flow in only one direction through the meter. When the bottom line of the input

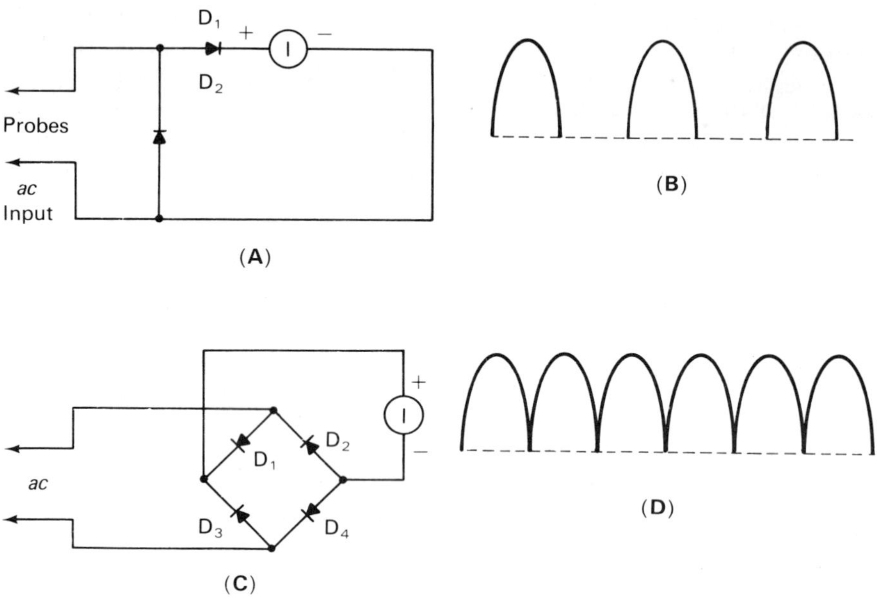

Figure 2-4 Rectifiers in ac meters.

probe is negative and the top positive, electron flow is from the negative terminal through the meter and rectifier and back to the positive upper terminal. When the bottom terminal is positive and the upper negative, however, the diode does not conduct, hence the meter receives a series of alternations of ac as shown in Fig. 2-4B. To prevent the reverse voltage from entering the meter circuit, an added diode (D_2) is used on occasion to shunt the reverse-polarity current.

The average current for each positive ac alternation (one-half of a cycle) produced by rectification is equivalent to 0.636 times the peak current. Since the circuit in Fig. 2-4A uses half-wave rectification, the average value is present only for half a cycle, hence the average over a complete cycle (two alternations) and for successive cycles is only half the 0.636 value. Thus, for a 100 mA peak current (with a root mean square of 70 mA) the average value would be $63.6/2 = 31.8$ mA. When the meter is used to measure voltages the same factors prevail and the scale must be calibrated accordingly so that root mean square values are indicated. (The root mean square value of ac voltage or current is equal to the amount of work the same current and voltage would perform if dc values were used.)

For better performance, full-wave instead of half-wave rectification is preferable. With full-wave rectification both alternations of the ac are used and the average value is then 0.636 of the peak value without the halving of the final value. A common method for obtaining full-wave rectification in meters is the bridge-rectifier circuit shown in Fig. 2-4C. Four diodes are needed, with two used to process each alternation of the ac cycle. With full-wave rectification the negative alternations of the ac are inverted as shown in Fig. 2-4D, thus producing single-direction current flow.

For the circuit in Fig. 2-4C, assume the top input line is positive and the bottom negative. Under this condition electrons flow to the junction of the two diodes D_3 and D_4. Only D_4 is wired to pass a negative alternation, hence electrons flow through it and through the meter to the junction of D_1 and D_3. Since the positive top line must be the return for the electron flow, they pass through diode D_1 and back to the source. Similarly, when the top line is negative and the bottom positive, electrons flow through D_2, through the meter and return to the bottom input line via D_3.

With full-wave rectification, the meter is calibrated to read root mean square (effective) values by designing the meter dial to read approximately 1.1 times the average value. Thus, a sinewave with a peak amplitude of 100 V has an rms value of about 70 V and an average value of 63.6 V. Hence, when the meter reads an ac sinewave having a 100 V peak value, the meter scale is graduated so the resultant 63.6 V is multiplied by 1.1 for a reading of 70 V rms. Similarly, for a current meter reading a peak amplitude of 100 mA, calibration would result in a reading of the rms value of 70 mA.

The frequency of the ac tested is also a factor. Servicing meters calibrated in rms and used on the 60 Hz supplied in the U.S. will give inaccurate results for audio signals having frequencies other than 60 Hz (because of reactance changes in the moving coil of the meter). Similarly, incorrect readings will be obtained for testing amplitudes of pulse waveforms or composite sinewave signals consisting of a number of other signal components.

2-5 CLAMP-AROUND AC METER

The reading of high currents in ac circuitry presents problems because of the necessity of opening the circuit to insert the ordinary ammeter. When heavy conductors are present, opening the circuit is usually impractical and special devices must be used for current readings. One procedure would be to read the voltage drop across the load circuit. Knowing the resistance of the load then permits calculation of the current, since $I = E/R$. Another method is to use a special clamp-around ac meter of the type illustrated in Fig. 2-5.

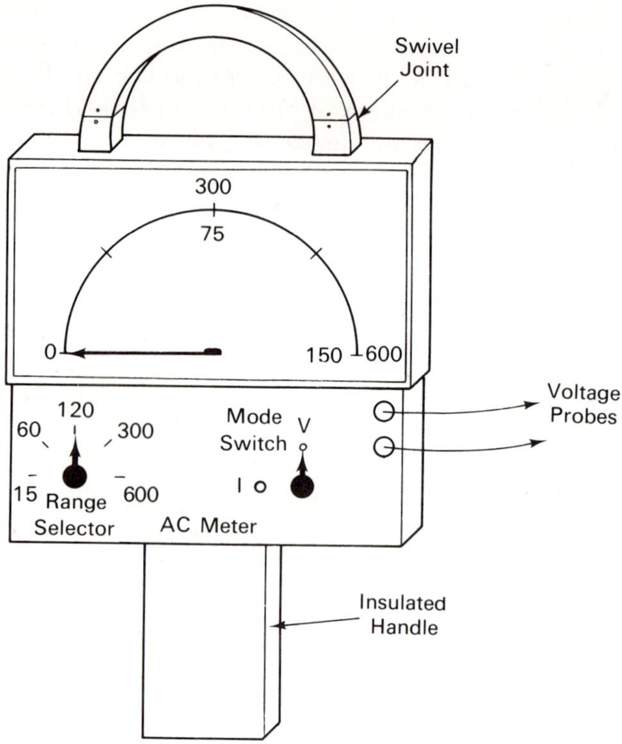

Figure 2-5 Clamp-around ac meter.

This meter reads alternating currents without the necessity for either opening the circuit or reading voltages across a known resistance. The basic meter movement is of the D'Arsonval type with a full-wave bridge rectifier such as shown in Fig. 2-4C.

The unit contains a split-core transformer, with the primary winding forming a one-turn loop that can be swiveled open to hook around a single conductor of the ac line. Because current flow through a wire creates magnetic lines of force around the wire, the alternating current flow and fields induce a voltage in the primary single-turn loop. Such a voltage drop is induced into the secondary winding of the meter transformer. Because the induced voltage is proportional to the current flow in the conductor, the meter scale is calibrated to show current readings. Two external terminals are also supplied so the instrument can be used for voltage readings of ac when needed.

The clamp-around loop must be applied to a single conductor only. If applied to the two conductors of the ac line, current flow in one direction through one line and in the opposite direction in the other, would create opposing magnetic fields and cause improper operation.

2-6 THERMOCOUPLE AMMETER

Thermocouple meters are used for measuring amplitudes of signals having frequencies of the RF type extending to several megahertz. The thermo elements form a circuit for measurement utilizing the basic dc meter movement for rms readings. The thermo elements are available with glass enclosed (vacuum-tube) as well as open-air types. A typical vacuum glass-enclosed unit is that shown in Fig. 2-6.

When the test probes are placed in series with the circuit under test, current flows through the heater wire and its temperature rises. This increased temperature is transferred to the thermal junction composed of two dissimilar metal strips. When the junction of the two strips heats up, a potential difference is produced at the ends. The voltage is then measured by the meter, which is calibrated to read either units of voltage or current. For voltage readings a series resistor is usually present between the heater wire and the test probes.

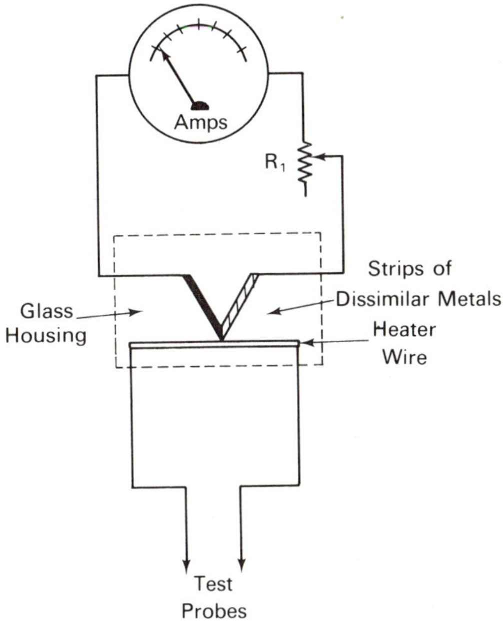

Figure 2-6 Circuit for thermocouple ammeter.

With a junction temperature at approximately 300° F, the voltage appearing at the ends of the thermocouple will average between 8 and 10 mV, depending on the type unit in operation. The unit is calibrated by adjustment of resistor R_1. The meter reading is directly proportional to the temperature of the heater wire, which is proportional to the square of the current through the resistance of the heater. Thus, the device measures rms current values in terms of I^2R. The vacuum thermoelements in some units are operated up to approximately 1 A, while the air coupled tubes are often used for higher current calibrations.

2-7 CHECKING CIRCUIT CURRENTS

In general testing, servicing or adjustment procedures involving radios, television receivers and high-fidelity stereo or quadraphonic devices the technician rarely needs to read current values. Since the circuit must be opened it is usually more convenient to make other tests during routine servicing procedures, including ohmmeter readings of components, voltmeter checks at various points in the circuit, and an inspection of signals viewed on an oscilloscope screen. When current readings are indicated, an open circuit must be formed for insertion of the current-reading meter, or the voltage drop across a known resistor must be read and current calculated on the basis of E/R.

Where the manufacturer designates current readings to be taken, provisions may be included in the circuit for convenience in applying the current meter. One such instance occurs in some television receivers (color) that use an adjustable horizontal efficiency coil for peaking circuit performance after replacement of a tube or component in the horizontal output section of the receiver. As shown in Fig. 2-7A, a removable cathode-circuit jumper is provided so that cathode current can be read while adjusting the horizontal efficiency coil tunable core. Thus, current may read 175 to 200 mA and the core is adjusted for a dip in current reading (to approximately 150 mA or as recommended in the service notes for the receiver). At the current dip the circuit efficiency is at a maximum for best performance.

When the currents drawn from a power supply are to be read, a convenient method is to remove the fuse as shown in Fig. 2-7B and apply the current-reading meter to the fuse terminals. Thus, the open circuit is closed by the current meter and the amount of current drawn by the filter system of the supply and the load are indicated. Of course, meter polarities must be observed so the needle pointer does not move in the wrong direction.

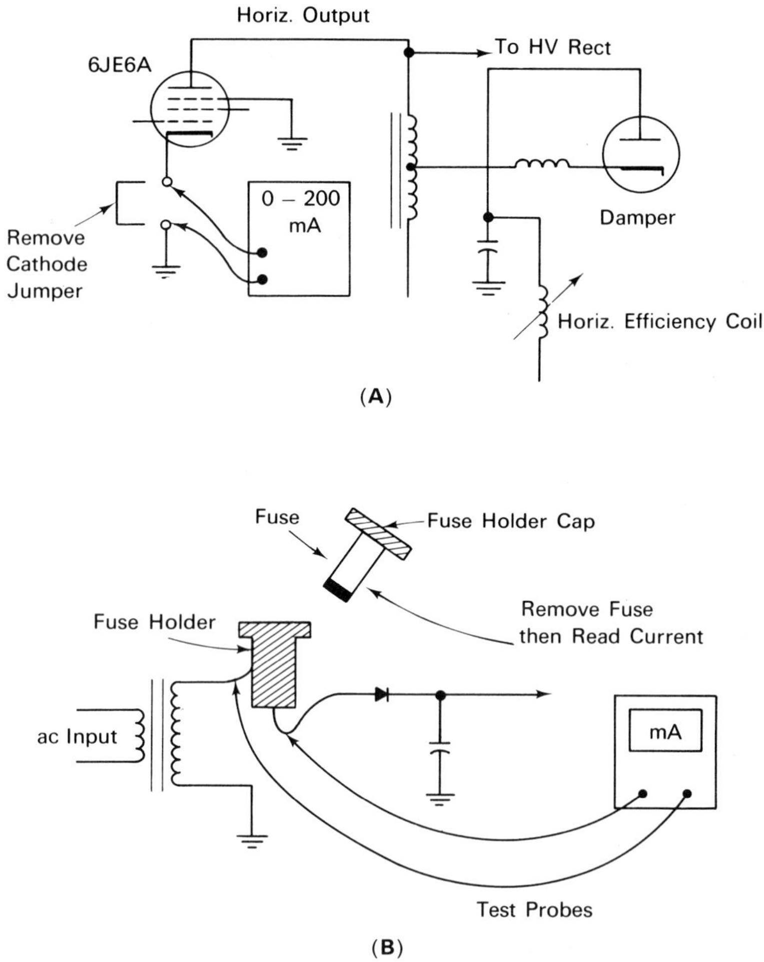

Figure 2-7 Circuit current checks.

2-8 BASIC OHMMETER

An ohmmeter, as the name implies, is a device for measuring the ohmic value of circuit components. Thus the values of resistors can be determined, as can the ohmic dc resistance of coils and other components. Basic milliammeters or microammeters are the units used in the design of ohmmeters and the more sensitive meters permit the reading of higher ohmic values.

The basic ohmmeter circuit is shown in Fig. 2-8. Two resistors are generally used in a series circuit with a battery and one side of the

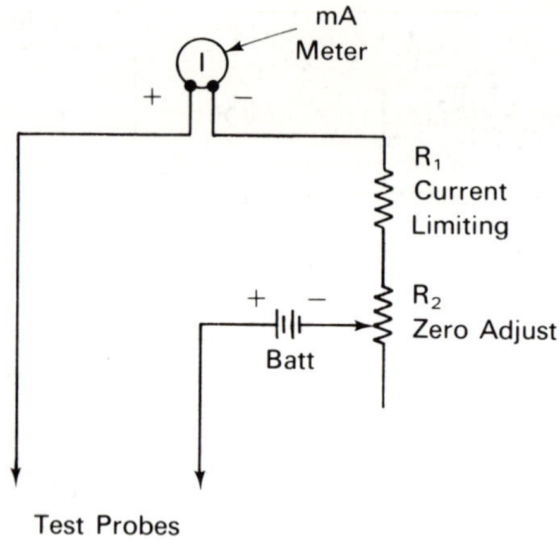

Figure 2-8 Basic ohmmeter.

meter as shown. Resistor R_1 limits the current flow through the meter within the range in usage and R_2 adjusts current flow so that full-scale deflection is obtained when the test probes are shorted together. When the pointer is at full-scale deflection (with probes touching each other) *zero resistance* is indicated, hence the dial is calibrated to read from right (zero resistance) to the left (maximum resistance). This results in a nonlinear scale with crowding at the high end at the left, as seen in Fig. 2-9.

The basic circuit of the ohmmeter functions on the voltmeter principle as more fully described in Chapter 3. Thus, the two resistors must have a total combined resistance to limit current flow through the

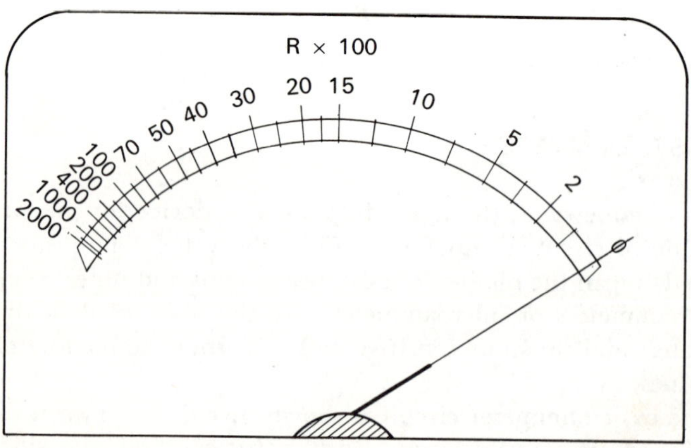

Figure 2-9 Typical ohmmeter scale.

meter to that required for full deflection (with R_2 at about half setting to permit zero adjustments). If the meter had an internal resistance of 100 Ω, and full-scale deflection is achieved at 0.1 V, we would need a 3-V battery if the resistors had a combined value of 2900 Ω.

Ohmmeters are usually combined with voltmeters and current-reading meters to form volt-ohm meters (VOM) and similar test instruments as covered more fully later. With FET units and highly sensitive meters readings in the megohms are possible. Also, with the more sensitive instruments (such as the 50 μA or 100 μA meters) lower values of battery potentials may be used, thus minimizing battery costs in portable equipment. (See also Sec. 5-2, bridge-circuit meters.)

When measuring the ohmic value of resistors and other units, it is preferable to disconnect one side of the unit from the circuit to avoid the shunting effects of other parallel resistances. In most circuit design, some variations in resistance readings are permitted (10 percent to 20 percent in some instances). Thus, a reading of 900 to 1100 Ω for a 1000-Ω resistor may be satisfactory in a number of instances. (The service notes for the device under test should be consulted to ascertain the allowed tolerances of resistors, coils and so forth.)

2-9 CONTINUITY TESTING

An important servicing and testing procedure is that of *continuity checking*. The ohmmeter is particularly useful for this purpose; its resistance-reading characteristics will indicate whether an open or short circuit exists. In continuity checking, however, the ohmic value of components is not usually considered—only whether circuit continuity occurs and whether open or short circuits prevail.

As an example, assume we wish to check a power supply of the type shown in Fig. 2-10. Initially the switch is opened, though it would be better to remove the plug entirely from the wall socket to avoid shock hazards as well as prevent damage to the ohmmeter. With the plug removed a continuity check can be made (with switch open) to ascertain if a short circuit exists. If no reading is obtained, the switch is closed and the meter-needle should deflect, indicating a closed primary circuit. These procedures fulfill the continuity-checking requirements, though if transformer overheating occurs (or if low voltages are present during operation) there is a possibility of shorted turns. If this is suspected the actual ohmic value of the primary must be read and compared to the service notes for the device, or as otherwise indicated by the manufacturer (see Sec. 1-7).

Again, with power shut off continuity checks can be continued on the secondary side of the transformer as shown in Fig. 2-10. Continuity for each side of the transformer (as against ground) should be made. A continuity check at the output of the rectifier diodes will indicate whether the filter capacitors are partially shorted and are creating an abnormal load because of the leakage. Open bleeder voltage-divider resistors are also checked for continuity (R_2 and R_3 as shown).

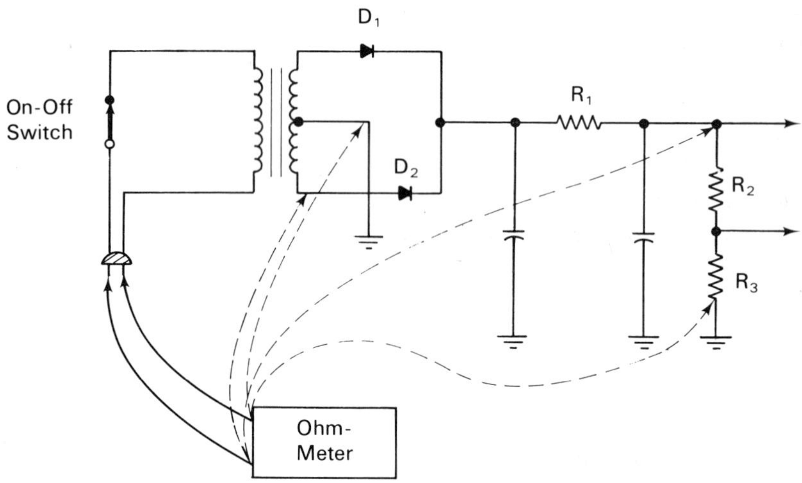

Figure 2-10 Continuity checks on power supply.

Similar continuity checks are performed on any other circuits requiring trouble shooting. If, for instance, an RF amplifier such as shown in Fig. 2-11 requires servicing, power is again shut off and continuity checks undertaken at various points in the circuit. A check across L_2 will show whether an open-coil condition exists. (With RF coils a short-circuit condition may be incorrectly assumed because of the low resistance of the coil. A short circuit is rare, however, and the continuity checks are for open circuits.)

A similar check across L_4 will establish the continuity for this inductor. Next, a reading can be undertaken from the junction of C_3 and L_3 to the bottom of L_5 as shown, to see that continuity exists between these points. When continuity checking FET units (as opposed to junction transistors) a low-voltage ohmmeter (on its highest resistance range) is preferred, particularly for the MOSFET devices (see Chapter 8, Sec. 8-5).

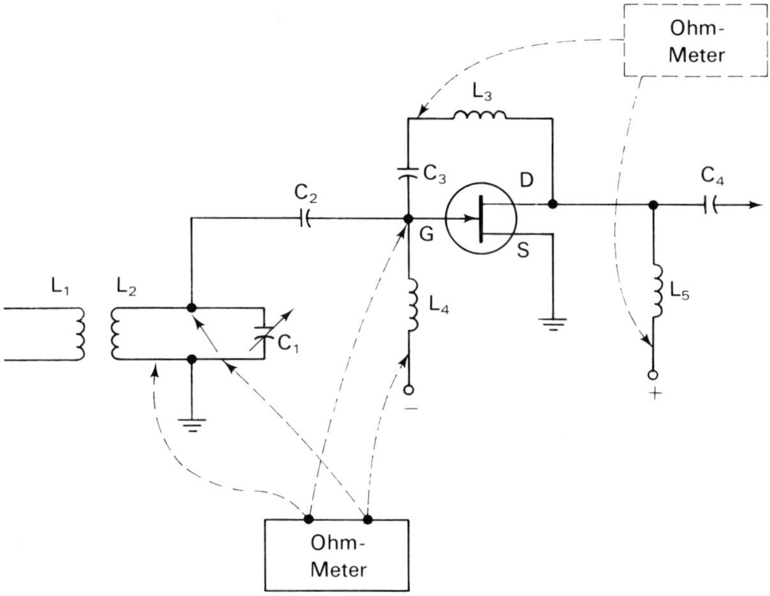

Figure 2-11 Continuity checks on RF amplifier.

3

voltmeter types and applications

3-1 THE BASIC VOLTMETER

The less-expensive voltmeters in common usage employ the D'Arsonval meter movement (see Sec. 2-1) while the more elaborate types use digital displays, as will be more fully described later. Voltmeters are available in single units or are combined with ohmmeter and milliameter units (VOM) with multiple ranges for each. The type utilized depends on requirements and cost factors.

The voltmeter measures voltage drop *across* a resistor or other component in contrast to the series measurement used for current readings. When an ammeter is put in series with current flow it must have a low resistance so that it will not consume any of the current that flows. In contrast, the voltmeter must have a high resistance so that it does not upset circuit function by decreasing circuit resistance. If the voltmeter loads down the circuit it will consume electric energy and upset voltage readings.

Thus, if the voltage across a 30 kΩ resistor is to be measured, the internal resistance of the voltmeter should be considerably higher than 30 kΩ because the voltmeter resistance parallels the resistance across which the voltage is measured. This is evident when one considers the loading effect that occurs if the voltmeter has the same resistance as the unit across which voltage is measured. If the meter resistance is 30 kΩ and the resistance across which the meter is placed is also 30 kΩ, the current will divide between the two 30 kΩ resistances and thus disturb circuit characteristics and give false voltage readings.

As with any current-reading meter, the basic voltage measuring meter is limited in the amount of current that can flow through it. Thus, most of the voltage drop that occurs across the voltmeter probes must be developed across a *series* resistor so that only a fractional voltage value appears across the meter coil. If, for instance, the basic meter is a 50 μA type with an internal resistance of 2 kΩ, the voltage drop across the meter coil at full-scale deflection would be $IR = 0.00005 \times 2000 = 0.1$ V. Assume we wish to increase this meter range to read a full scale of 5 V. To accomplish this it will be necessary to add a series resistor across which 4.9 V will drop so that this voltage plus the drop across the meter (0.1 V) adds to 5 V. If we permitted more than 0.1 V to appear at the meter it would force more than 50 μA of current through the coil and damage it.

For this meter the external resistor would be $R = E/I = 4.9/0.0005 = 98$kΩ. Equation (3-1) simplifies the resistance calculation:

$$R_s = R_m(N - 1) \tag{3-1}$$

where R_s is the ohmic value of the series multiplier resistor.
R_m is the internal resistance of the meter coil.
N is the number by which the meter voltage scale is to be increased.

Thus, applying Equation 3-1 to the foregoing example, the number by which the meter voltage scale is to be increased is from 0.1 V to 5 V, or a total of $5/0.1 = 50$.

$$R_s = 2000\ (50 - 1) = 2000 \times 49 = 98{,}000\ \Omega$$

Thus, the 2 kΩ meter resistance plus the external resistor of 98 kΩ gives a total resistance of 100 kΩ with 50 μA of current flow through it. Hence we have our 5 V full-scale deflection as $100{,}000 \times 0.00005 = 5$ V.

3-2 OHMS-PER-VOLT SENSITIVITY

Voltmeter sensitivity is rated in *ohms per volt* and it denotes the ohmic value by which the multiplier resistor must be raised to increase the meter scale by one volt. Thus, voltmeter sensitivity can be considered to represent the ratio of the multiplier plus meter coil resistance to the voltage deflection of the meter. If a 50 μA meter has a sensitivity of 20,000 Ω per V, as compared to only 1000 Ω per V for a less sensitive type, the former would have much less loading effect on the circuits being measured.

The ohms-per-volt sensitivity can be found by initially using Equation 3-1 to find the series resistor ohmic value needed to get a 1 V full-scale reading. Thus, increasing the 0.1 V meter to 1 V means that the number by which the voltage scale is to be multiplied is 1/0.1 = 10, hence:

$$R_s = 2000\ (10 - 1) = 2000 \times 9 = 18{,}000\ \Omega$$

For the same meter, an increase to a 2-V full-scale deflection means the proportionate increase is 2/0.1 = 20 for the number by which the meter scale is to be increased. Thus, we obtain

$$R_s = 2000\ (20 - 1) = 2000 \times 19 = 38{,}000\ \Omega$$

Comparing the results obtained for the 1-V and 2-V full-scale deflection we find that an increase of 1 volt in full-scale deflection required a 20,000 Ω increase in the multiplier resistance (from 18,000 Ω to 38,000 Ω), or 20,000 Ω per V. Thus, the meter sensitivity for this 50 μA meter is 20 kΩ per V. To increase the full-scale deflection by another volt (to 3 V) would require an additional 20 kΩ of series resistance. Thus, such a meter used on the 100-V full-deflection scale would have a multiplier resistance of 1998 kΩ. Thus, on this scale, the meter resistance (including the 2 kΩ internal resistance of the coil) would be 2 MΩ and would have a negligible loading effect on most circuits being tested. With 100 μA meters with still higher internal meter resistance, greater sensitivities are achieved and thus precision measurements may be made.

3-3 VOM UNITS

When voltmeters are combined with ohmmeter and current-reading meter functions, appropriate switching of shunt and series resistors, plus ohmmeter battery, must be used to select the meter operation needed. Since all three operations may have multiple ranges, the switching circuitry can become somewhat complex. The voltmeter alone, for instance, may have three or four ranges on the dc scale only.

If the voltmeter has three ranges the series resistors could have the circuitry shown in Fig. 3-1A or B. In Fig. 3-1A a selector switch engages the desired series resistor, which provides for the full-scale deflection needed. Depending on the meter design, this could be a full-deflection 3-V scale, a 30-V scale or a 300-V scale. The arrangement shown in Fig. 3-1B could be used in which resistors R_2 and R_3 combine their individual resistance with the other series resistors. Thus, in the upper switch position only R_1 is in the circuit. In the middle switch position, the total multiplier resistance is the combined ohmic values of R_1 and R_2. In the lower switch position, all three resistors combine to provide the required multiplier resistance.

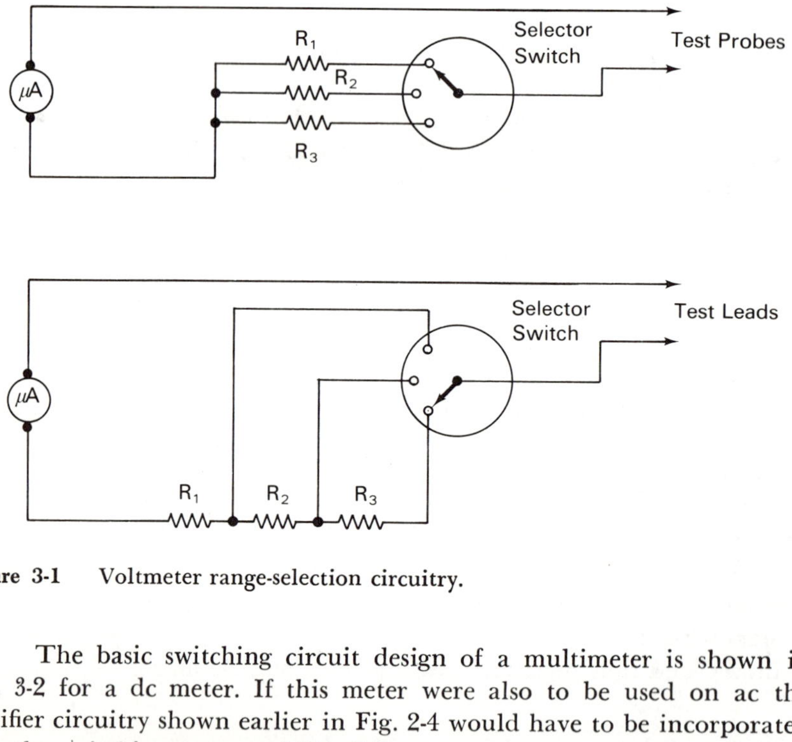

Figure 3-1 Voltmeter range-selection circuitry.

The basic switching circuit design of a multimeter is shown in Fig. 3-2 for a dc meter. If this meter were also to be used on ac the rectifier circuitry shown earlier in Fig. 2-4 would have to be incorporated into the switching system of Fig. 3-2.

For the current-reading scales, four ranges are provided: 100 mA, 10 mA, 1 mA and 50 μA. The latter position does not involve any shunt resistors because the basic meter range is utilized. For the other current scales, resistors R_6, R_7 and R_8 are placed in shunt as required. For voltage readings, resistors R_3, R_4 and R_5 are selected individually and placed in series with the meter movement to multiply the voltage-reading scale. Ranges for this particular meter thus furnish readings of 10 V, 50 V and 500 V dc.

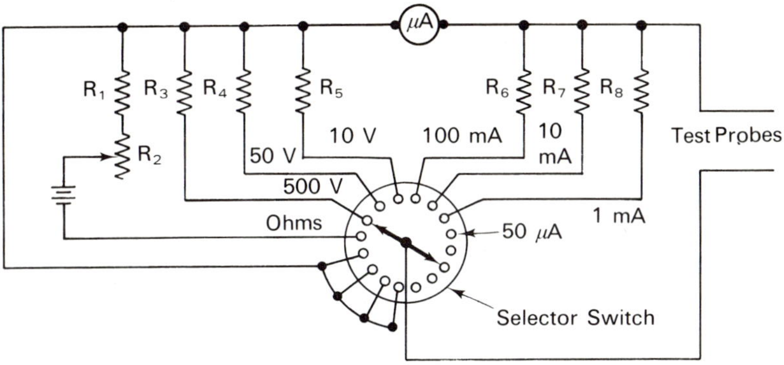

Figure 3-2 Basic circuit of multimeter.

Although the front panel layouts for commercial meters differ from each other, a typical VOM panel with representative function selections is shown in Fig. 3-3. Voltage ranges (either dc or ac) are from 0.5 V to 1000 V as shown. For this particular meter push-button selection is made for either the ac or dc mode of operation. The push-button marked +/− reverses the polarity input of the probes for reading negative values.

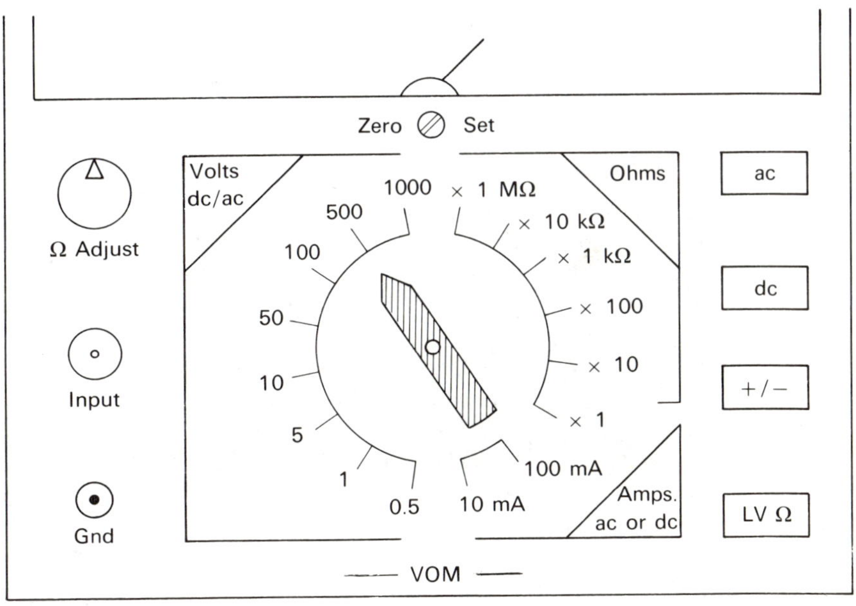

Figure 3-3 Typical VOM panel.

The ohmmeter range multipliers are shown at the right. For the x1 selection, the meter scale is read directly for the ohmic value being measured. For reading higher ohmic values, the various selector positions permit multiplication of the meter scale by 10, 100, 1,000 or 1,000,000 as shown. Thus, the maximum amount of resistance that can be measured depends on the sensitivity of the meter used (see Sections 2-8 and 3-2).

A low-voltage ohmmeter selection is also possible so that transistors and other solid-state devices can be checked without voltage overload and possible damage to the units. With the low-voltage selection the meter applies less than 0.1 V to the circuit under test.

Two current-reading selections are present, though with most commercial meters four or five current ranges can often be selected. With some meters only dc can be measured, while the more versatile types permit measurement of both direct and alternating currents.

Most commercial VOM units have available special probes for extending the usefulness of the meter. RF probes are available that permit measurement as high as 50 MHz. In addition, high-voltage probes can be plugged into the VOM for reading voltages as high as 50 kV. The high-voltage probe thus permits direct reading of the 25 kV or higher voltages found in color television receivers. The probes have insulated handles and high-voltage insulating fins as shown in Fig. 3-4. The elongated probe places the hand safely away from the high-voltage source (and also contains the high-voltage, high-resistance, voltage-dropping units).

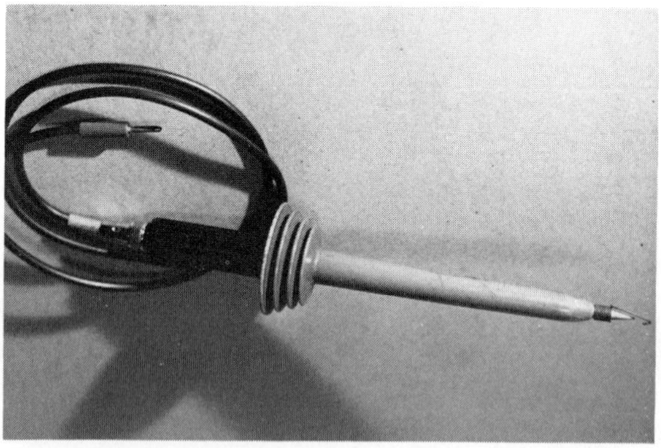

Figure 3-4 High-voltage probe.

3-4 VTVM OR FET VOLTMETERS

For years one of the most popular VOM units was one employing the rectifying-amplification advantages of tubes. By channeling the input to the grid of the tube and sensing the degree of amplification, the circuit under test was subjected to a minimum of loading and the meter exhibited increased sensitivity and versatility. Such a meter was known as a vacuum-tube voltmeter (VTVM). As the state of the art advanced, solid-state units took over the function of the tube. Consequently, transistors, silicon diodes and MOSFET units are widely used in modern test equipment, with the field-effect transistor (FET) forming a VOM superior to the older VTVM (see also Chapter 8).

With the older VTVM units, only the dc scales had the high-impedance feature so desirable in these meters. Also, the VTVM did not have provisions for current readings, since 1 mA meters were generally used and these were insensitive for wide-range current scales. Also, the VTVM suffered the drawbacks of warm-up time and (compared to solid-state) more frequent tube troubles.

The FET meters overcome many of the drawbacks of the VTVM types. No warm-up time is needed, aging has no appreciable effect on performance, thus insuring accuracy for longer periods than with the VTVM. Also, the FET units operate on extremely low current, which assures economy and long battery life.

As with the VTVM, the FET units have a high input impedance, with some FET units running as high as 15 MΩ on dc. (Most VTVM units had an input *Z* of approximately 10 MΩ.) Also, in contrast to the VTVM, the FET units provide for current measurements, since meters with basic movements of from 50 to 100 μA are used.

3-5 DIGITAL METERS

Where a high degree of accuracy is desired, digital meters are useful, though their cost is much greater than that of the VOM or FET meters. Digital VOM units are available which read ac or dc with selector-switch convenience, as shown in Fig. 3-3. The display is in numerals, as in Fig. 3-5A. Digital display units are used to form the readout, hence the numerals are light displays against a black (or dark-colored) background.

Some of these meters feature automatic sensing and display of the polarity, preceding the reading with a + or − and read to three places to the right of the decimal point. Accuracy is often from +0.1% to +0.5% of the reading, + 1 digit.

Small hand-held digital VOM units are also available with a general configuration as shown in Fig. 3-5B. Some have voltage ranges up to 1000 V dc and read resistance to 20 MΩ. Input resistance may exceed 50 MΩ for a minimum of circuit loading when in use. With the digital

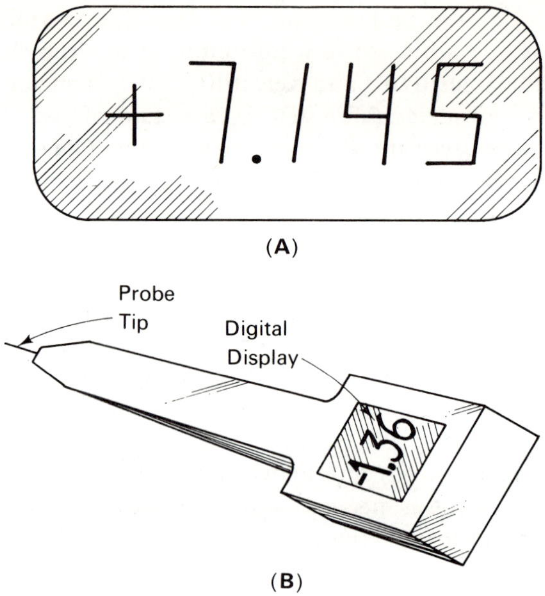

Figure 3-5 Digital meter displays.

meters the reading is obtained as soon as the probe is applied (less than 3 s), automatically displaying polarity and decimal point (such as +12.44, −1.332, −864.2, +9.521).

3-6 DECIBEL SCALES

Most VOM units have a scale for measuring decibels. The *bel* (named after Alexander Graham Bell) is not used. Instead, one-tenth bell (decibel) is the standard used as a comparison between powers, voltages or currents. A decibel (dB) change in sound represents the increase in sound intensity barely distinguishable to the average person. Doubling power equals 3 dB. Decibels are additive, that is, if we increase the power output from an amplifier from 2 watts to 4 watts, we have a +3 dB change. Increasing the power again to 8 watts (double the 4 watts) adds another 3 dB, thus giving us a change of 6 dB from the original 2 W to 8 W. A doubling of current or voltage, however, is equal to 6 dB. Thus, increasing the signal voltage output from an amplifier from 0.2 mW to 0.4 mW will equal 6 dB.

The ear responds to a change in sound intensity in a fashion that is logarithmic. Thus, the ear is much more responsive to changes in low-intensity sound levels than to higher-intensity levels. Mathematically, the decibel for power is based on logs:

$$dB = 10 \log_{10} \frac{P_1}{P_2} \tag{3-2}$$

When the voltage (or current) in a circuit is *doubled,* power does not double, but *quadruples.* Hence the equations for E and I calculations of dB are as follows:

$$dB = 20 \log_{10} \frac{E_1}{E_2} \tag{3-3}$$

$$dB = 20 \log_{10} \frac{I_1}{I_2} \tag{3-4}$$

Reference levels are often used to simplify analysis and measurement of power, voltage and current changes, such as 0.006 W (6 mW) across 500 Ω. Additional data on this will be found later in this section regarding meter scales.

Nepers is another term encountered and it expresses the same factors as dB. The dB is common to the United States; the neper has been used extensively in Europe. The neper uses the natural base of logs, ϵ (2.7182. . . .). There is a constant relationship between the dB and neper (n), where 1 dB = 0.1151 n, and 1 n = 8.686 dB.

In equation form we can solve for n by:

$$n = \frac{1}{2} \log_{\epsilon} \frac{P_1}{P_2} \tag{3-5}$$

Most VOM units include a dB scale on the meter face. A typical representation is shown in Fig. 3-6. Here the *zero* dB is at 0.7746 V on the voltage scale, and the dB scale may be read directly provided the measurements are made across a 600-Ω load resistance. The dB scale represents *power* ratios and many meters use the reference where zero dB is 1 mW of power across a 600-Ω load resistance. This then represents the 0.7746 V of ac *(rms)* across the 600-Ω load. (The dB scale is directly related to one of the ac scales of the meter, usually the lowest-value one.) The abbreviation *dBm* is used to designate dB based on 0 dB = 1 mW, 600 Ω.

Some meters have used a reference based on zero dB = 6 mW at 500 Ω, where the 0 dB point is at 1.732 V on the voltage scale. Another reference has been 0 dB = 6 mW, 600 Ω, dB point at 1.897 V. In all cases measurements provide for direct readings when taken across the

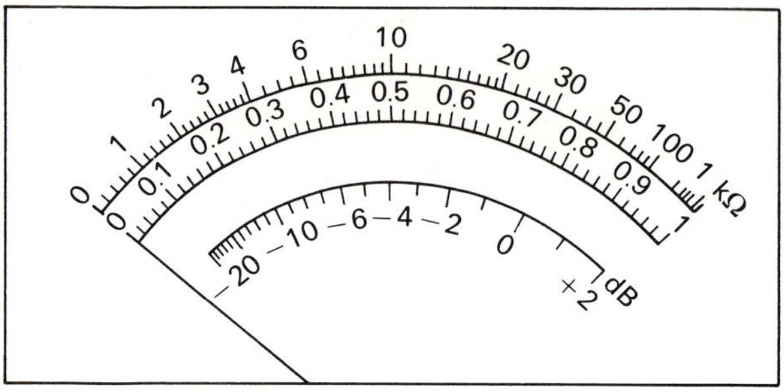

Figure 3-6 FET VOM with dB scale.

load resistance recommended. Thus, if the output from one amplifier is measured across 600 Ω and indicates 6 dB and the output from another is also measured across a 600-Ω resistance and reads 10 dB, we have a 4 dB difference (a little over double the power output). Had the resistances each been 400 Ω instead of 600, the relative dB value would still be correct and would be 4 dB, though the absolute dB readings obtained would not be true values.

When measurements are made to compare the dB difference between the input and output signals of an amplifier or other unit, the ohmic values of the resistances across which measurements are taken must be identical to avoid incorrect readings. For dB measurements, test signals should be of the sinewave type. For measurement of signals having complex waveshapes such as square waves, or audio-type multifrequency signals, refer to VU discussions in Sec. 3-7, which follows.

3-7 VU (VOLUME UNIT) METERS

The *volume unit* (VU) is widely used in audio testing and measurement and describes a decibel-oriented unit. Tape and cassette recorders usually have two VU meters for monitoring recording levels to prevent overload and distortion. With the VU, the zero level is assumed to equal 0.001 W across 600 Ω of Z. In equation form we obtain:

$$\text{VU} = 30 \log_{10} P \qquad (3\text{-}6)$$

A typical VU meter dial is shown in Fig. 3-7. The scale extends from −20 to +3 VU on the upper range. The lower scale is marked in *percentage* of voltage, with 100 per cent situated at the 0 VU point. For most meters the numbers and scale markings to the left of 0 are black, and to the right they are red. Normal volume level is considered

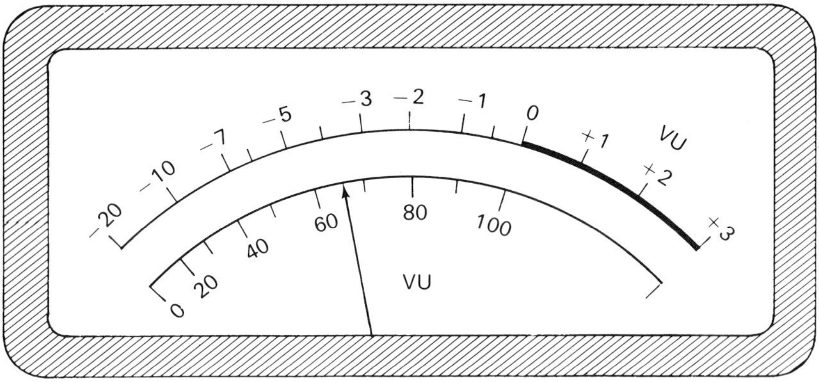

Figure 3-7 VU meter dial.

to be at the 0 VU point (100 per cent), which is located at about 70 percent of full-scale deflection.

The VU meter movement uses full-wave rectification and hence reads an approximate *rms* value of the complex ac waveform of audio signals. When in use the VU meter should be observed over a short period so that several peaks of the signals can be observed and the VU level adjusted with controls to maintain normal limits and avoid overloading.

The *dBm* mentioned in Sec. 3-6 refers to the number of dB above or below the 1 mW reference level, with sinewave signals and resistive loads. The sinewave test signals are maintained at a constant amplitude, hence *dBm* is used only for single-frequency constant-amplitude sinewave signals. The VU measurement, however, applies to signals of varying frequencies and amplitudes with instantaneous peaks often exceeding the 1 mW *dBm* reference. VU readings can be taken across an impedance (combination of resistance and reactance) with the average power much less than the 1 mW reference for *dBm*. Thus, VU is actually an arbitrary value and has no fixed value as is the case with the *dBm* described in Sec. 5-6.

3-8 PRACTICAL VOLTAGE TESTS

An inoperative television, radio, hi-fi system or appliance can be caused by defective components, blown fuses, defective line cords or lack of voltage from the wall outlet or other source. Since component testing or continuity checks would be meaningless if the voltage source has been cut off the initial tests should be for voltage input. If the unit uses line voltage from the ac mains the line cord should be removed and the ac voltage checked as shown in Fig. 3-8A. Lack of voltage or voltages below normal call for a check of the wiring, fuses, circuit breakers and so forth.

If they do not deliver the proper charging potential battery chargers can drain batteries instead of charge them. Chargers must furnish slightly more voltage than the normal battery rating, so that when connected the electrons from the negative lead of the charger are forced into the negative side of the battery to replenish the charge. Return electron flow is from the positive terminal of the battery to the positive terminal of the charger. A simple voltage check as shown in Fig. 3-8B will establish whether the correct potentials are present. If not, ascertain that good contact is being made by the charger prongs to the wall outlet. Also check the charger's outlet cord for breaks. With the charger removed from the wall socket, a continuity test (Sec. 2-9) will determine whether an open circuit exists.

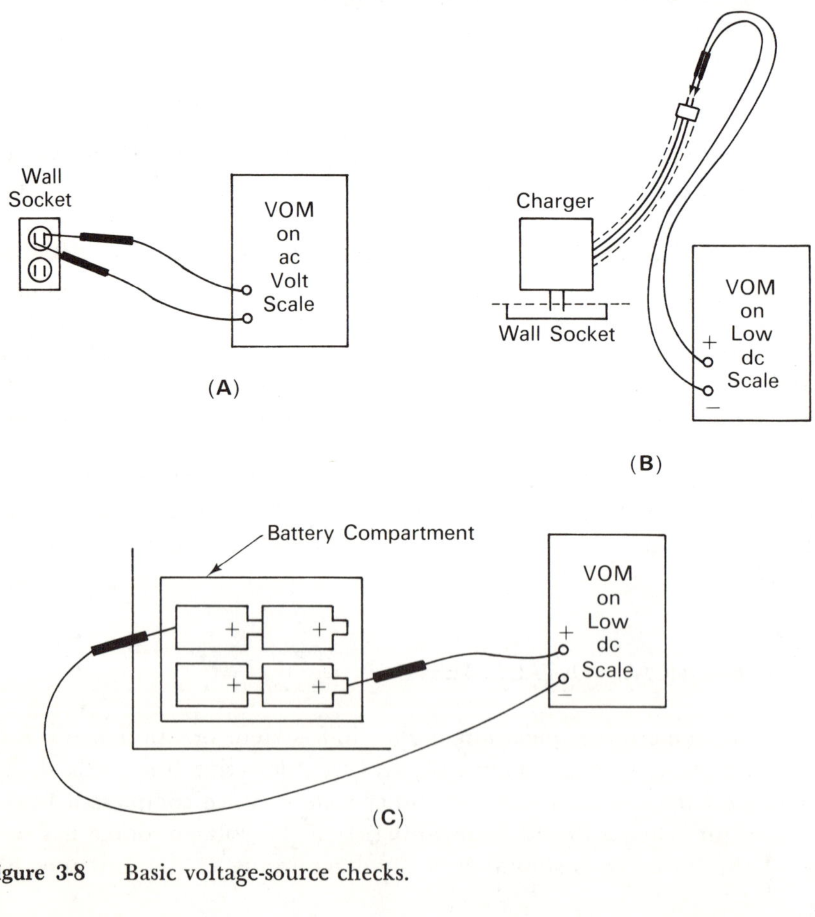

Figure 3-8 Basic voltage-source checks.

Weak batteries, of course, will also contribute to poor performance in portable radios, cassette players and other equipment. If the batteries are removed and tested they may check out as having near normal voltage. However, when they are in use in the equipment they are working under load conditions and battery potentials drop considerably if the battery has been in use for some time. Thus, battery voltages should be measured while the batteries are operating the equipment. With the radio or other unit turned on, open the battery compartment and note the number of batteries and the type connections used. Batteries in series must provide voltages near the sum of the potentials involved. Thus, if four cells are used as shown in Fig. 3-8C, the voltage should be near the 4 × 1.5 value, or 6 V. Probably the appliance will still operate at 5 V or less under load conditions, but with decreased output. When the series voltage drops below the 5 V the unit may become inoperative. The test probes, however, must be placed at the most negative and positive terminals for correct readings.

If line voltages or battery packs have proper available potentials but devices are still inoperative, circuit tests will have to be made. Tests on power supplies should involve the points shown in Fig. 3-9A. Using the ac scale the voltmeter should be placed across the transformer primary (test points 1 and 2) to check proper operation of the switch and to make certain that the input fuse, if used, has not blown. If voltages are present here, the meter is set on the dc scale that is high enough to prevent meter overload. The negative probe of the meter is placed at ground (test point 7) and the positive probe is placed at test point 6 to find out whether the power supply has a voltage output. If it does, voltage and component checks will have to be made on other circuitry in the unit.

If no voltage is present at the output (test point 6) the positive probe can be placed at test point 5. If voltage is present here but not at point 6, the series resistor has opened. If no voltage is present at point 5, the meter will have to be reset to the ac scale and readings taken at test points 3 and 4. Voltage at point 3 but not at 4 indicates a blown fuse. Voltage at point 4 but not at 5 means a defective rectifier is present.

The sequence of such tests can be varied to suit the technician. After finishing the ac test at point 1 and 2, for instance, the meter could be left on the ac scale and moved to points 3 and 7 or 4 and 7, then changed to a dc setting to test points 5 and 6. Abnormally low voltages could be caused by leaky filter capacitors or an excessive load on the supply caused by shorts or partial shorts in other circuitry that uses power from the supply.

Voltage checks at the input and output circuitry of tubes and transistors are useful to measure circuit performance, localize defective

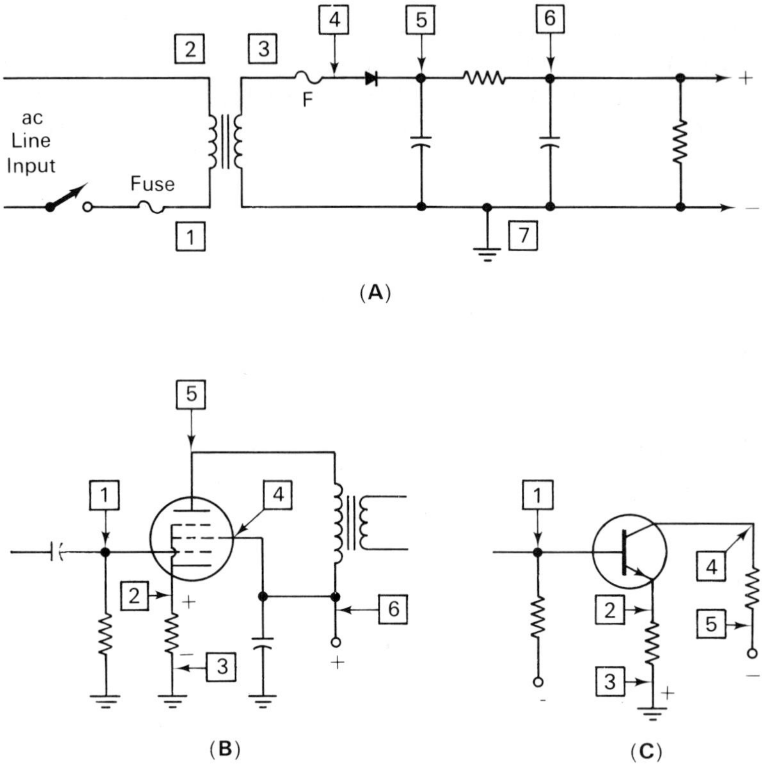

Figure 3-9 Circuit voltage checks.

parts or find points of abnormal voltages. Various check points for a pentode tube circuit are shown in Fig. 3-9B. Using the dc scale, with the negative probe at point 1 and the positive probe at point 2, the amount of tube bias can be measured. Thus, a reading of 1.5 V would indicate that the grid is minus with respect to the cathode by 1.5 V. Since the resistor in series with the cathode establishes the bias relationship between cathode and grid, the bias can be measured across this resistor also. For such measurement, however, the negative probe is placed at point 3 (ground) and the positive at the cathode input.

For Class A operation of amplifiers, the grid of the tube does not run positive (as with Class AB_2 or Class C), hence there should be no reading across the input grid resistor (from point 1 to point 3) because the grid does not draw current and no voltage drop should occur across the grid resistor. If a positive voltage is read between points 1 and 3 it indicates either a leaking input coupling capacitor in the grid line or a tube that has developed some internal gas and draws grid current.

Anode voltage is measured between point 5 and point 3, where a high positive voltage should be read. Similarly, the screen-grid voltage is measured between point 4 and ground. A below-normal voltage at the screen grid may be caused by a leakage in the screen by-pass capacitor. Absence of voltage at point 5 could be caused by an open wire connection or an open in the primary of the transformer. If no voltage is present at point 6, however, the power supply and feed lines to point 6 must be checked.

Similar tests are performed on transistor circuits, with typical test points illustrated in Fig. 3-9C. The bias between the base (at point 1) and the emitter (at point 2) should be negative at the base with respect to the emitter for a pnp transistor (see Chapter 8). For testing the voltage across the emitter resistor, the negative probe must be at the emitter with the positive one at ground (point 3).

For the npn transistor, the collector has a negative voltage applied to it and this is read with the negative meter probe at 4 and the positive probe at point 3. If no voltage is present, check at point 5. If voltage is read at point 5, it indicates an open series resistor or open circuit in the wiring from the resistor to the collector.

3-9 HIGH-VOLTAGE TESTS

Use of the high-voltage probe for testing the voltage at the second anode of a picture tube is shown in Fig. 3-10. The high-voltage input plug can be removed from the wall of the picture tube and the presence or absence of voltage checked. The plug should be reinserted into the tube wall socket, however, and the voltage read again to make sure load conditions do not decrease the applied voltage. If voltage decreases to below normal, the series resistor in the high-voltage line should be checked, as well as the high-voltage circuitry (see Chapters 10 and 11).

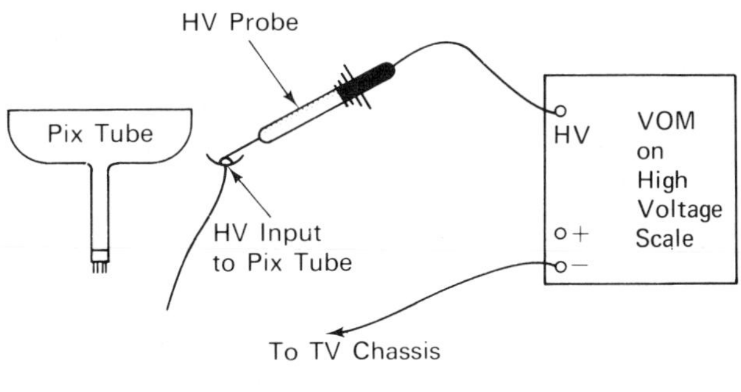

Figure 3-10 Testing TV high voltage.

When reading high voltages, precautions must be taken to avoid shock hazards. The ground connection should be hooked into place and one hand used for the high-voltage probe. Some probes have a U-section of wire attached to the tip to permit hanging the probe to the test point, leaving the hands free for making voltage adjustments and corrections. The manufacturer's notes or schematics should be consulted for the amplitude of voltage that should exist in the circuitry. Also, it must be remembered that high-voltage probes are designed for a specific instrument and are not interchangeable among various meters. Using the probe with meters other than the one for which it was designed results in incorrect voltage readings.

3-10 COLOR DETECTOR TESTS

Schematics also show other voltages that are normal for proper operation of the television, radio, high-fi system or other unit. Thus, for proper voltage tests a schematic of the unit under test is a necessity. Not only are voltages designated but also tolerances are given for resistor values, capacitor ratings and so forth. Usually voltages are designated as read by a high-impedance voltmeter with the receiver or other unit operating and controls set at normal. In most instances measurements are taken as against the common chassis (ground). A typical section of a television schematic is shown in Fig. 3-11 to illustrate the method of voltage designations. Here the B-Y and R-Y demodulators of a color receiver are shown. Note the specific voltage designations given. At the collector of the B-Y color detector we should read 6.5 V and if a 5 percent tolerance is designated, any of the voltages below 6.2 V or above 6.8 V call for testing associated components that may cause voltage differences.

Note that the collector for the R-Y demodulator has a different voltage on it than the B-Y detector. Each emitter has the same voltage, however, and again these must be within the tolerances specified by the manufacturer. Each base input should have approximately 23.2 V present as measured against ground, with 23.5 V present at the input feed line as shown. Often, as transistors and components age, the overall efficiency of the system declines. In such instances voltages that approach the tolerance limit (without exceeding them) may contribute to inferior performance. Thus, it may be advisable to bring the voltages closer to those specified to restore peak performance.

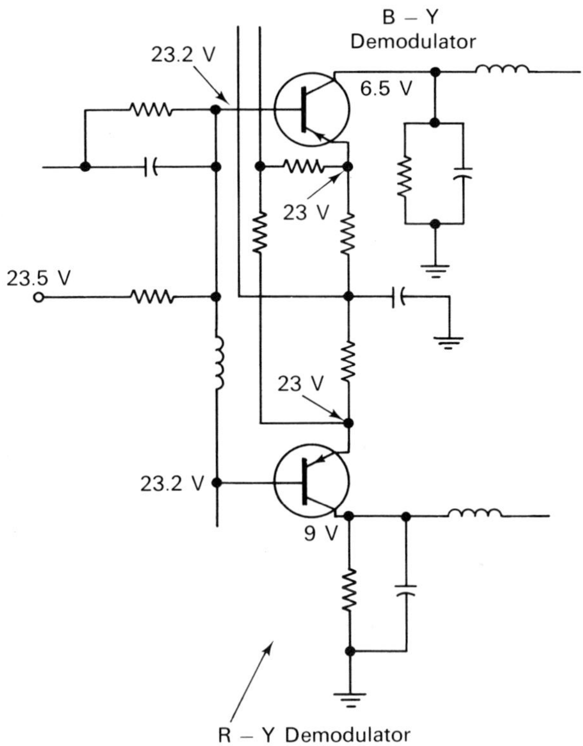

Figure 3-11 Color detector voltages.

3-11 AVC AND AGC CHECKS

Automatic volume control (AVC) systems tend to maintain a constant volume-level output from radios once the volume control has been adjusted to suit the individual listener. A typical circuit is shown in Fig. 3-12. Note that the AVC system (R_1 and C_3) taps the AM diode detector circuit. The AM detector rectifies the incoming radio frequency signal (the IF signal in this instance) and filters out the RF by C_2. Thus, C_2 bypasses RF signals from across R_2 (the volume control) but does not shunt the audio. Hence, the output to the audio amplifier consists of ac-type signals that vary in frequency and amplitude.

Resistor R_1 obtains signals from the AM detector system but bypasses the audio with C_3, leaving only the average value of the dc component. Because this average value increases for a stronger station, a greater AVC potential is obtained for an increase in the AM carrier signal arriving at the detector. The AVC potential is applied to the

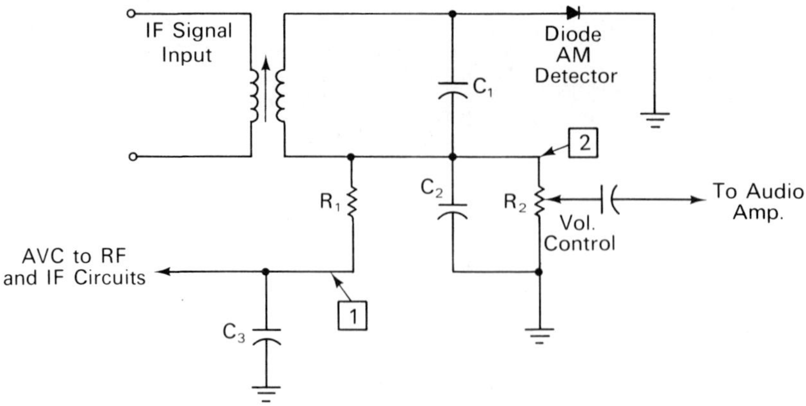

Figure 3-12 Detector and AVC circuit.

input circuits of the tuner RF stages and the IF stages, adjusting their bias and hence altering gain. Thus, the larger AVC obtained for a stronger station decreases gain in the early stages and compensates for the volume level rise that would otherwise result.

The AVC system's voltage can be read from test point 1 to ground with a low dc voltage obtained here. As various stations are tuned in, this voltage should vary accordingly. Failure to obtain a voltage here would call for a check of R_1 or a shorted C_3 capacitor.

Failure of the diode detector circuit would also affect the AVC system. Trouble with the detector, however, would cause decreased sound or total loss of audio. In that instance the diode should be checked to make sure it has unidirectional current characteristics. The signal voltages can be checked across test point 2, but for satisfactory tests at this point a scope is preferred (scope tests are detailed in Chapter 4).

Automatic gain control (AGC) systems are similar in operation to AVC, except that AGC systems are used in television receivers to maintain a constant video level once the contrast control has been set. A two-transistor AGC system is shown in Fig. 3-13, with the first transistor acting as a gate to turn on and off a second transistor. The latter (AGC amplifier) raises the AGC signal level to the amplitude needed for application to the input circuits of the RF and IF stages.

The AGC systems are usually of the *keyed* type; that is, a horizontal pulse is used to provide for open-gate conditions. Note that the AGC gate base input is from the video amplifier. These signals contain sync and blanking as well as video signals. It is only during the sync intervals, however, that the AGC gate conducts, because only at this time does it receive the necessary pulse input from the horizontal section.

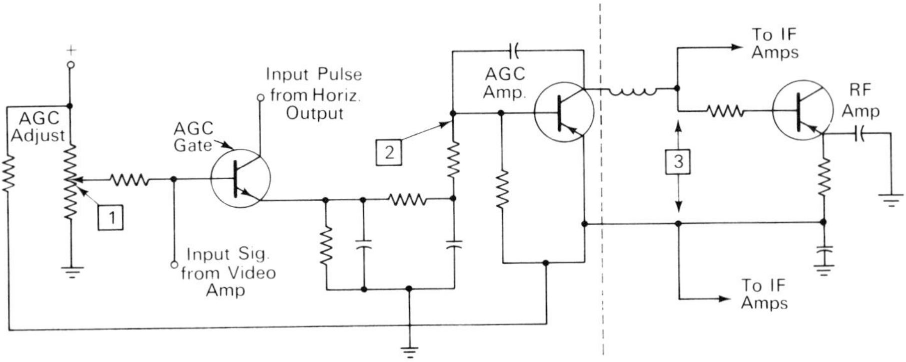

Figure 3-13 AGC system.

The pulse, in the form of a short burst of dc, provides the necessary collector bias for conduction. When the gate transistor conducts, it supplies bias to the AGC amplifier, which, in turn, develops an output dc that represents the AGC potentials applied to the RF and IF stages. This system eliminates the instability that would occur if video signals entered the AGC system and altered the output from the amplifier.

Picture bending, excessive contrast, or an almost dark screen may be caused by improper AGC settings. Thus, the AGC adjusting potentiometer is set for the strongest television signal in the area to prevent circuit overloading. Once set for the strongest station, the AGC system will maintain a fairly constant signal level except for extremely weak stations where signal strength is insufficient even when the AGC permits maximum gain in the RF and IF stages.

AGC troubles call for a check of the voltage at test point 1. Here the fixed bias (positive potential for the forward-bias condition) is set by the variable resistor and as this resistor is adjusted, the voltage input to the base should vary. If the gate circuitry checks out satisfactorily, a test can be made at point 2 to note whether the gate is furnishing the necessary forward bias to the base of the AGC amplifier. If no bias is present, check for possible shorts of the filter capacitors between the gate emitter and the amplifier base, as well as for a possible open series resistor. If this circuit tests normally the voltage can be tested at check point 3. As various stations are tuned in, the voltage arriving at the RF and IF inputs (between base and emitter) should vary.

If voltages are absent, check for possible defective transistors and associated components. Continuity testing (Sec. 2-9) will disclose shorted or open circuitry.

4

oscilloscope applications

4-1 USEFULNESS OF SCOPE

The oscilloscope is also referred to as *the scope* among technicians and engineers. It is an instrument that uses a cathode-ray tube for displaying pictures of various signals used in electric and electronic circuitry. Thus it provides a visual indication of possible troubles in signal reproduction or amplification. Also, it permits a display of the bandpass characteristics of RF and IF stages, thus facilitating alignment and tracking procedures in television, frequency-modulation and other receivers. Similarly, the displayed image can be utilized for signal-tracing purposes as described more fully in Chapter 6.

Signals to be displayed are usually applied to the scope input, which initiates deflection in a vertical plane and displays amplitude changes of the signal. This feature is also useful when the scope is used as a peak-to-peak reading voltmeter as described later in this chapter. An internal sawtooth-forming oscillator scans the input signal horizontally and forms the time-base factor of the displayed signal. By incorporating special facilities into the oscilloscope circuitry, frequency measurements can be made and the type of display more rigorously regulated as described later.

The basic structure of the oscilloscope tube is shown in Fig. 4-1. Electrostatic deflection is accomplished by applying voltages to the horizontal and vertical deflection plates to influence the position of the electron beam. The beam has a negative charge and is attracted by a positive field and repelled by a negative one.

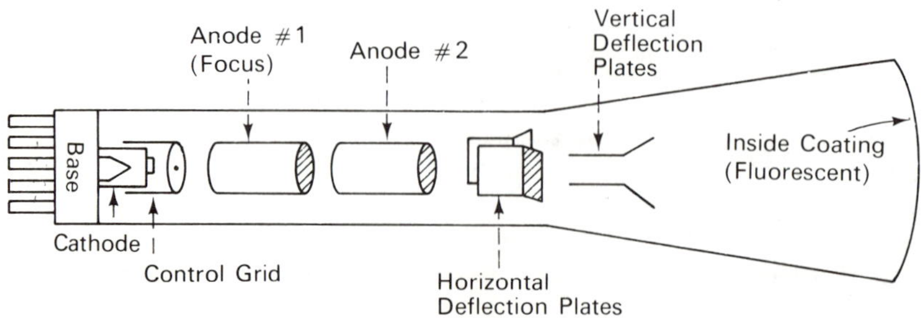

Figure 4-1 Typical oscilloscope tube using electrostatic beam deflection.

The cathode of the tube is heated by filament wires and emits electrons. The electron stream from the cathode is regulated by the control grid structure so its intensity can be adjusted as required. Anodes one and two have positive voltages applied on them to attract and accelerate the electrons and form them into a beam. The voltage applied to the first anode is made variable to permit focussing the beam into a sharp pinpoint of light. The two anodes are comparable to a lens system because they focus the beam for good definition at the tube face. The assembly that forms the electrons into a fast-moving beam is termed an *electron gun*. Additional grid and gun structures may be used in special-purpose tubes and dual-trace types described later.

Besides the signal voltages applied to the deflection plates, a controlled amount of dc is applied to both the vertical and horizontal plates for adjusting the initial position of the electron stream as it impinges on the tube face (position control).

4-2 FUNCTION OF CONTROLS

The basic controls on the front panel of an oscilloscope are shown in Fig. 4-2. Their placement on the panel may vary from one manufacturer to the other, but functions remain the same. The intensity control adjusts the bias on the control grid to vary the brilliance of the displayed

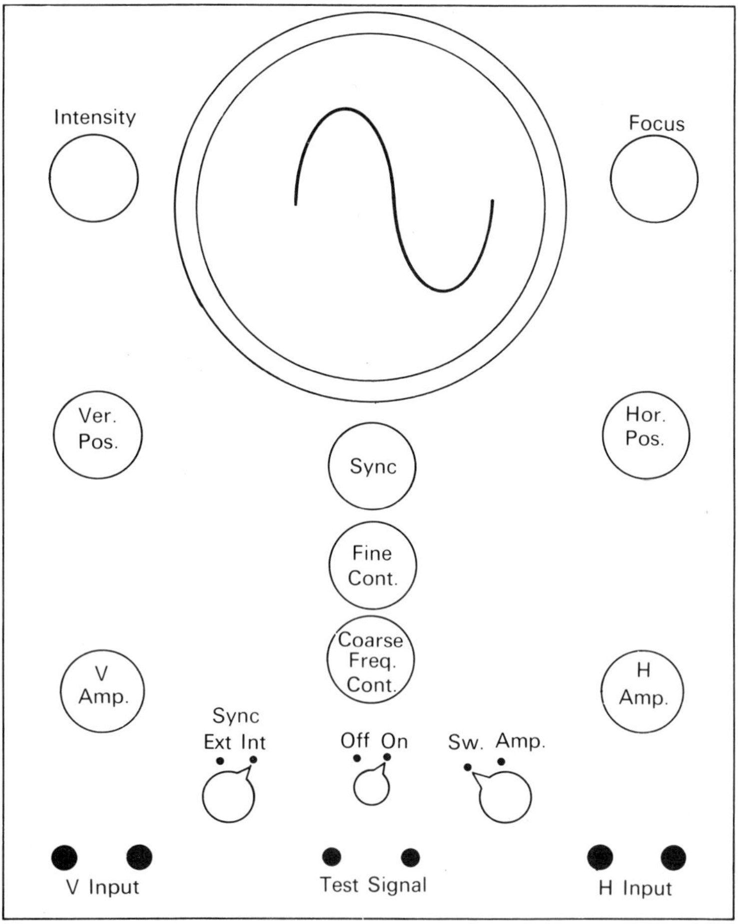

Figure 4-2 Typical scope controls.

image. The focus control adjusts voltages on the anode section of the cathode-ray gun to pinpoint the beam trace. The oscilloscope circuitry contains a signal generator (sawtooth oscillator) used to trace the image across the tube face. The sawtooth signal is applied to the horizontal plates and as the right-hand plate becomes more positive than the left plate, the electron stream is attracted to it. When the sawtooth incline drops suddenly to zero the beam retraces to the left to start the scan over again.

A switch is available on the panel to shut off the horizontal sweep so an external sweep can be applied to the horizontal input amplifiers. A synchronizing switch also is used to select either internal or external synchronization as needed. Variable resistors form the horizontal sawtooth signal frequency selectors and sync control. A fine-tuning control is often available for precise adjustment of internal sweep frequency.

The usefulness of the scope depends on the technician's familiarity with the function of the controls and the inherent possibilities of the scope. As shown in Fig. 4-3, the signal displayed on the screen can be shifted horizontally or vertically, or expanded in either plane to provide the type display needed for convenient analysis of the waveform.

4-3 UNDERSTANDING DISPLAYED SIGNALS

When the internal sawtooth sweep signal of the oscilloscope is switched off and a sinewave signal applied to the horizontal input as well as the vertical input, phase and frequency determinations can be made. The sinewave signal for the horizontal input can be obtained from a signal generator calibrated for accuracy of the frequency read on the dial.

When both the vertical and horizontal input signals have the same amplitude, the pattern changes indicate the changing phase relationships as shown in Fig. 4-4. These are known as *Lissajous* patterns, after the French scientist who first observed them experimentally. If both input signals have a stable frequency, any of the patterns shown can be locked into synchronization for a stationary pattern display.

A typical example of a 45° phase difference is shown in Fig. 4-5. A difference in amplitude between the two signals will elongate or otherwise alter the displayed pattern. Failure of the two signals to synchronize properly will cause a continual shifting through the patterns shown in Fig. 4-4.

When one of the input sinewave signals is twice the frequency (or one-half the frequency) of the other a figure-eight pattern is produced as shown in Fig. 4-6. Thus, one signal could be 60 Hz and the other 120 Hz, or one could be 400 kHz and the other 200 kHz. A 3 to 1 ratio with respect to the frequency of the input signals will produce the pattern shown in Fig. 4-7. For successively greater frequency ratios more of the elongated circle patterns will appear. If the signals to the vertical and horizontal inputs are reversed the pattern assumes a vertical position, though the number of elongated circles that appear still indicates the frequency ratio between the two applied input signals.

For general testing and troubleshooting procedures the internal horizontal sawtooth sweep signals are used and adjustments made to synchronize the sawtooth with the input signal to display one cycle or more of the signal under observation. Although a dozen or more sinewave or other signals could be displayed on the screen, it is preferable to lock

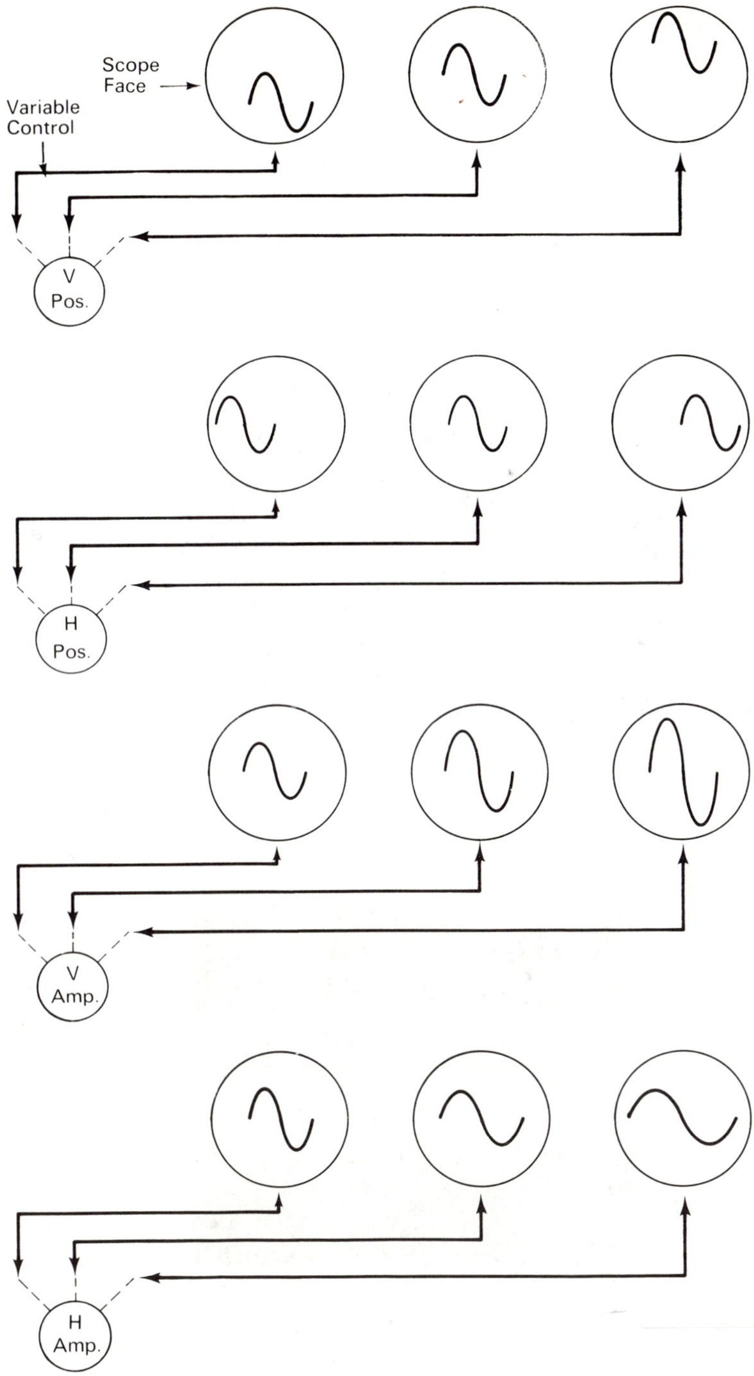

Figure 4-3 Function of scope controls.

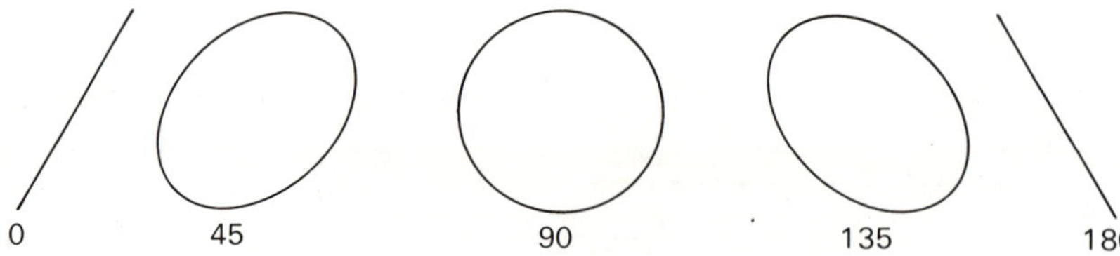

Figure 4-4 Lissajous patterns showing phases.

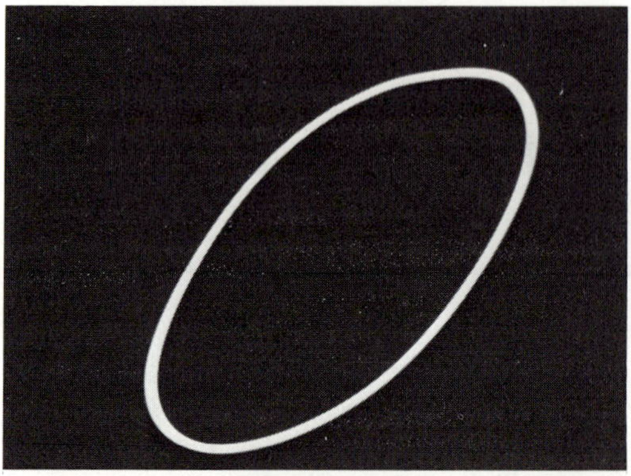

Figure 4-5 Signals at 45° phase difference.

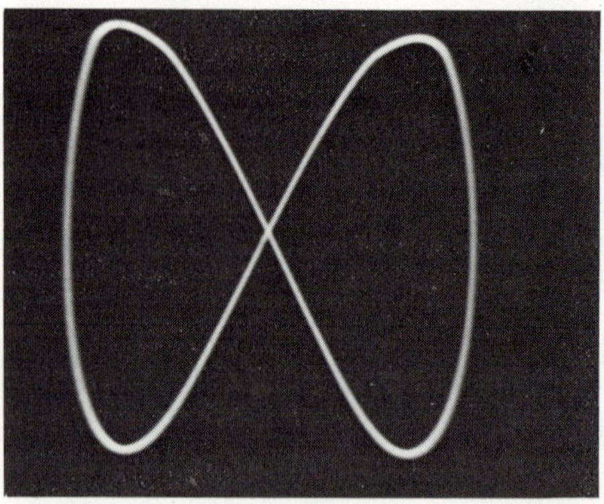

Figure 4-6 Signals having a 2/1 frequency ratio.

in only one cycle or pulse waveform for better observation. A jumble of signals as shown in Fig. 4-8 indicates that there is no synchronization between the input signal and the internal sweep signal of the oscilloscope.

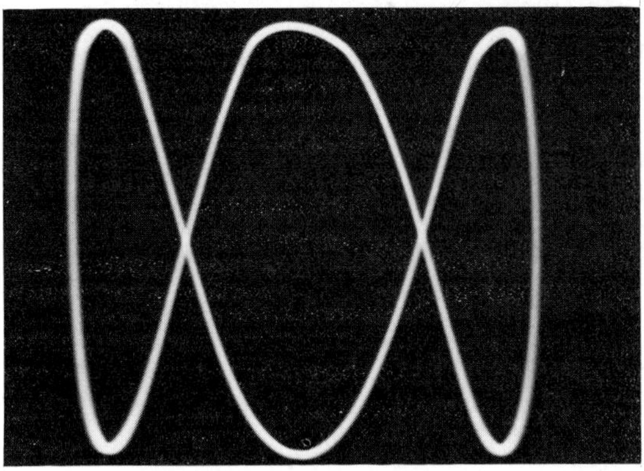

Figure 4-7 Input signals with 3/1 frequency ratio.

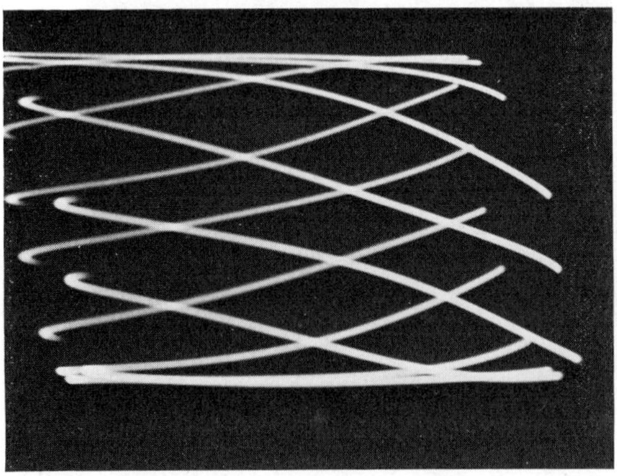

Figure 4-8 Input signal and sweep signal not synchronized.

A pure sinewave signal is shown in Fig. 4-9 and is characterized by a smooth rise and fall of amplitude and a symmetrical relationship between successive alternations (half cycles). When such a signal is applied to an amplifier (from a signal generator) it should pass through the amplifier without a change of form except for amplification. If, at the output, a signal such as shown in Fig. 4-10 appears, it represents distortion of the sinewave signal. Note that there is no longer a smooth transition to and from the peak amplitudes, since some flattening occurs. Such

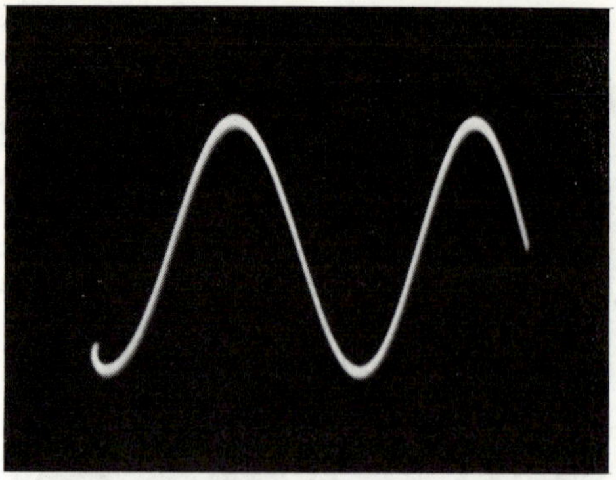

Figure 4-9 Pure sinewave.

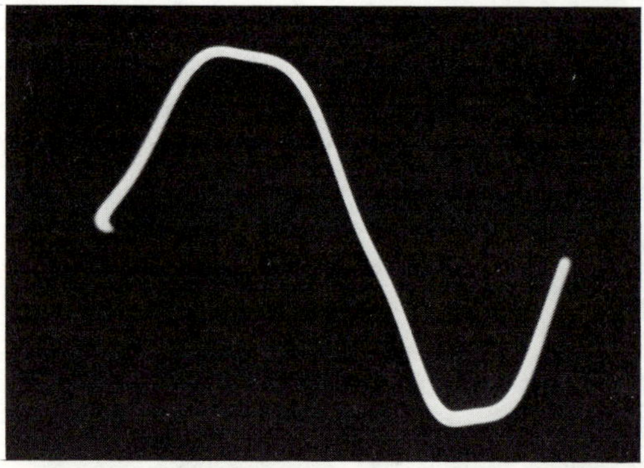

Figure 4-10 Distorted sinewave.

distortion indicates the presence of other signals (harmonics) in addition to the fundamental-frequency signal.

A square wave signal such as shown in Fig. 4-11 is preferable to the sinewave signal for testing distortion in amplifiers. The square-wave signal is made up of a number of odd harmonics, with each higher harmonic having a decreased amplitude. Thus, the fundamental-frequency signal, plus the third, fifth, seventh, etc., form the composite signal, which has a square waveshape as shown. The vertical sides represent high-frequency signal components, and the flat top represents lower frequency signal components. The first vertical incline is termed the *leading edge* and the decline vertical line is called the *trailing edge*.

Figure 4-11 Typical square wave.

A distorted square wave is shown in Fig. 4-12. Note that the leading and trailing edges have been retained (indicating presence of high-frequency signal components) but the flat-top portion now has a decided curvature indicating loss of low-frequency signal components. Such a waveshape is also called a *differentiated* waveform and in some instances is deliberately formed to perform certain functions such as switching and pulsing. If this waveform appears from an amplifier where a pure square wave has been injected, however, it indicates loss of low-frequency signals. A change in coupling capacitor value can cause loss of low-frequency components. Similarly, poor design where the coupling capacitor is too small in value (thus having a high reactance in series with low-frequency signals) can result in differentiation of the signal.

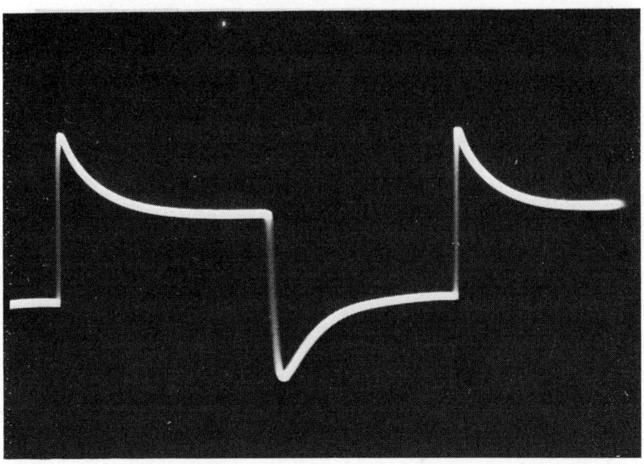

Figure 4-12 Differentiated waveform.

When the circuitry introduces loss of high-frequency signals the distorted square wave may appear as shown in Fig. 4-13. The curvature at the top of the leading edge and the bottom of the trailing edge shows high-frequency signal loss, thus reducing the amplitude of some of the high-frequency harmonics that contribute to the waveshape. Such a signal is also termed an *integrated* waveform and is sometimes deliberately formed (such as in the vertical sweep system of television receivers) for proper synchronization timing.

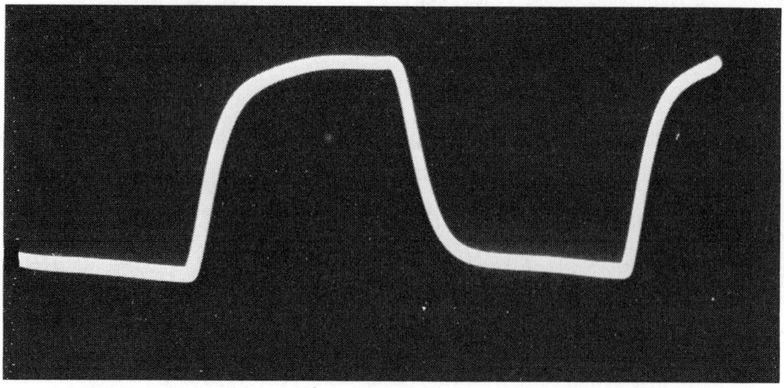

Figure 4-13 Integrated waveform.

In amplifiers or other electronic circuitry, the loss of high-frequency signal components may be caused by shunt capacitance between signal-carrying wires or components and chassis, or between wires (such as poor-grade coaxial cable or shielded microphone or phono cables). The integrated waveform output (when a pure square wave is applied to the input) also indicates that all high-frequency signals will be diminished as compared to the low-frequency signals. Thus, where a 400 Hz signal may undergo a gain of 50 through an amplifier, an accompanying 9 kHz signal may only have a gain of 30 or 35, depending on the degree of shunting effect for the higher-frequency signals.

Integrating and differentiating effects represent a distortion of the original square wave. Such differentiation and intergration do not appear when sinewave signals are injected into an amplifier and only amplitude changes occur (because a pure sinewave consists of a single frequency only, without higher harmonic content).

The mathematical factors relating to integrated and differentiated signals are somewhat complex and need not be considered for purposes of servicing and troubleshooting. The integrator circuit, when used, consists of a series resistor and shunt capacitor and can be considered as a low-pass filter (since some high-frequency signal components are shunted). The differentiating circuit, when used, contains a series capacitor and shunt resistor and can be considered a high-pass filter since lower-frequency signal components are diminished.

4-4 OSCILLOSCOPE TYPES

Many types of oscilloscopes are available commercially. The lower-priced ones perform basic display functions and the higher-priced models incorporate many features that facilitate waveform observation, analysis, testing and measurement. It is always best to select the scope that is the most versatile within the price range so as to expedite its usage.

As an illustration, some lower-priced oscilloscopes do not have retrace blanking, consequently retract lines will be visible on virtually all displays. The retrace line is not as bright as the primary trace, though still visible, as shown in Fig. 4-14. Compare this to the patterns shown in Figs. 4-9 through 4-13. Another fault shown in Fig. 4-14 is in the linearity of the horizontal trace. Note the second positive alternation of the cycle is not as wide as the first indicating a more rapid incline of the horizonal sawtooth within the scope. If internal readjustments of the scope's horizontal linearity are possible this condition may be corrected. If the scope has poor design little can be done to obtain a linear horizontal trace.

The better oscilloscopes also have an improved frequency response in the internal vertical amplifiers, thus permitting observation

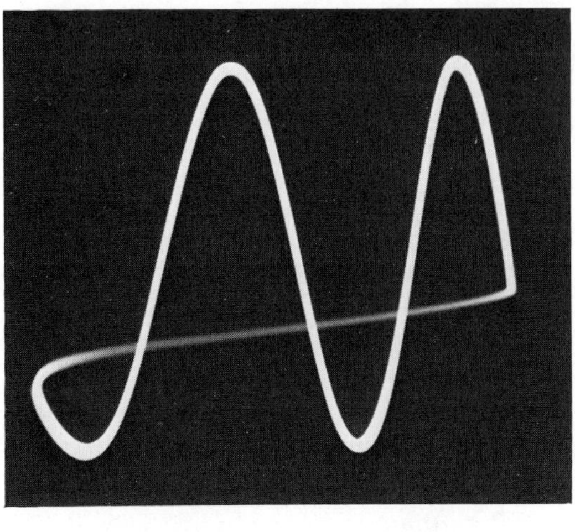

Figure 4-14 Scope pattern without retrace blanking.

of higher-frequency signals without signal distortion. It is difficult to interpret scope patterns if the vertical amplifier of the scope introduces distortion of signals by diminishing high-frequency signal components.

The high-frequency response required depends on the type of equipment with which the scope is to be used. If, for instance, the scope will be used for AM radio servicing, high-fidelity mono or stereo phono and tape amplifiers, and public-address systems, a scope with a response of from approximately 20 Hz to 30 kHz will be satisfactory. For television receiver testing purposes a frequency response of from 20 Hz to about 5 MHz is needed for visual display of the high-frequency signals involved. For research and design work involving VHF television signals, a scope may be needed that has a response in excess of 50 MHz for observation of video IF signals (see Appendix C and Chapter 10).

In addition to an adequate bandpass in the vertical amplifier of the scope, the display of high-frequency signals also depends on the range of sweep-frequency signals present. Only when the sweep frequencies are sufficiently high is it possible to display one or two cycles of a high-frequency signal.

A more elaborate scope for special testing is the dual-trace type which permits display of two different signals simultaneously. Dual-channel amplifiers and synchronizers are used for processing the individual signals and feeding them to the single cathode-ray tube. An electronic switching system alternately displays portions of each signal (usually displaced in the vertical position). The cathode-ray tube's phosphor-coated inner face has sufficient persistence to display the two signals properly. A typical display is shown in Fig. 4-15 where square waves are compared to sinewaves of the same frequency.

A dual-trace scope is useful in testing stereo equipment since it permits observation and comparison of the output of both channels. Any differences in waveform amplitudes or shapes can be observed more conveniently than by switching a single-trace scope back and forth between the two outputs. The dual trace scope is also useful for observation of phase differences between the 3.58 MHz demodulation signals in color receivers (oscillator and incoming burst). Comparisons can also be made between the signal applied to the input of a circuit and the signals appearing at the output.

Other methods may be employed to obtain dual trace. Two gun structures can be used for the cathode ray tube, though this increases the cost of the scope considerably. Also, an electronic switch circuit can be used for alternately channeling the signals into a single-trace scope. The disadvantage of this method is the necessity for using an external unit and the frequency limits imposed by many such devices.

An additional refinement in the more costly scopes is the feature

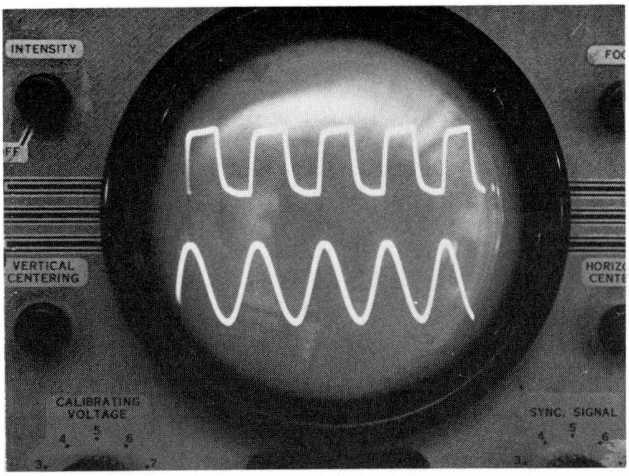

Figure 4-15 Dual trace permits waveform comparisons.

called *triggered sweep*. This sweep method provides for a more stable synchronization as well as a more precise one for it uses the input signal to initiate the trigger for horizontal sweep. In contrast, a scope without this feature uses a free-running oscillator to establish the sawtooth signal's frequency (also known as *recurrent sweep*). The free-running oscillator has a variable frequency control for selection of the necessary sweep signal time base. A sync circuit is used to lock in the sweep signal to the frequency of the signal to be displayed.

With the recurrent sweep a change in the amplitude or frequency of the input signal requires readjustment of the sync controls to stabilize the displayed signal. With the triggered sweep the lock in is automatic since the input signal is used to initiate the trigger that starts the horizontal sweep. A charging capacitor is used to create the rising sawtooth voltage that pulls the beam across the tube face. At the conclusion of the trace the sawtooth voltage suddenly drops to zero and the beam is retraced rapidly and awaits a new triggering pulse to set the sweep in motion.

In some scopes using triggered sweep, three switches are available for selection of the required mode, as shown in Fig. 14-16. Depressing the *free running* switch provides for recurrent sweep operation. In the *automatic* triggered mode a horizontal trace line is usually visible in the absence of an input signal. As soon as a signal is applied, the horizontal sweep is replaced by the triggered sweep mode. For the *manual* triggered mode the desired trigger point in relation to the input signal can be selected manually to permit examination of one area of the displayed signal.

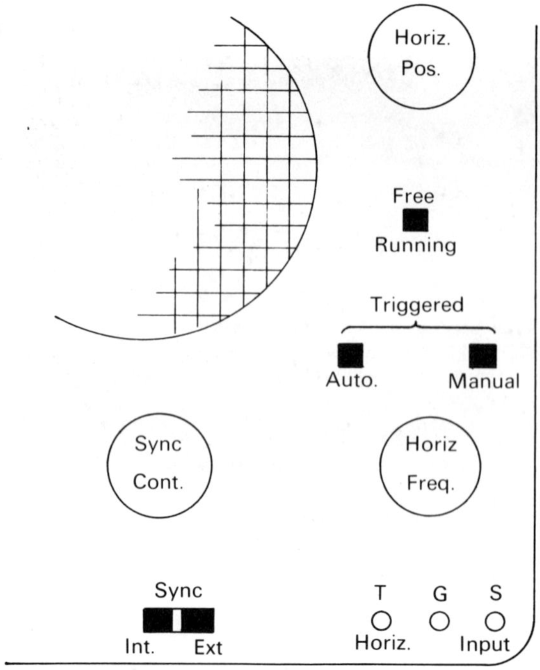

Figure 4-16 Trigger controls.

4-5 THE SCOPE AS A VOLTMETER

In testing and servicing television circuits and other electronic devices it is often necessary to ascertain the peak-to-peak (PP) voltage of signals present at certain portions of the circuitry. The scope easily can be calibrated to read such voltages and some models provide precalibrated controls for setting up the voltage-reading procedures.

For proper reading of peak-to-peak ac voltages, a transparent screen marked off with horizontal and vertical lines in graph formation must be used. This screen is placed over the face of the scope and becomes the reference for calibration to read voltage.

An ac voltmeter of known accuracy is used for calibration purposes. The voltmeter is used to measure the reference voltage (60 Hz) applied to the vertical input terminals. The internal horizontal sweep is shut off so that the trace on the screen will be a vertical line as shown in Fig. 4-17.

If the ac applied to the vertical input is 10 V, for instance, the resultant vertical line will have an amplitude depending on the setting of the vertical amplifier control. If the vertical control is adjusted to

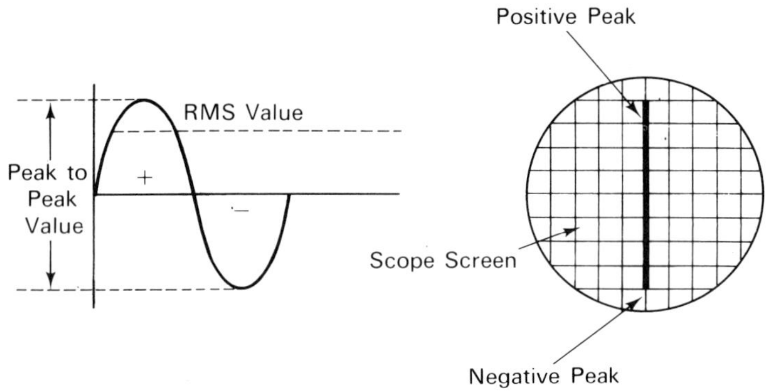

Figure 4-17 Peak-to-peak voltages on scope.

make the line trace two squares high, it would represent the peak-to-peak value of the 10 V input. The 10 V measured by the ac voltmeter, however, is the *rms* (root mean square) value (also called *effective* value, since it is the same as the dc value that would do the equivalent work). If the rms value is known, the peak can be found by the following calculation

$$\text{Peak value} = \text{rms} \times 1.41$$

Thus, the peak value of the 10-V rms value would be 14.1 V. Peak-to-peak voltage, however, would be twice this amount, or 28.2 V, as shown in Fig. 4-17. Now the scope will also read other voltages accurately so long as the vertical gain control is left untouched. Other values can be read to where the vertical line trace just fills the screen and calibration will hold for frequencies beyond 60 Hz with good accuracy (depending on the maintenance of a flat response in the vertical amplifier of the scope for the higher-frequency signals). In contrast, the ordinary ac voltmeter is calibrated only for 60 Hz and becomes increasingly inaccurate at the higher frequencies.

For reading higher voltages, the vertical gain is reduced and values recalibrated. Each square, for instance, could represent 30 volts, or any other value set by the calibration procedure.

Peak-to-peak (PP) voltage readings are often taken on television circuitry, particularly in the vertical and horizontal sweep sections where waveforms of 60 and 15,750 Hz are encountered, as well as 3.58 MHz burst in color receivers. Many schematics show scope patterns for trouble-shooting and also indicate the peak-to-peak voltages to expect. Audio voltages, too, are often tested for peak-to-peak values, as are square waves and pulses. Procedures are covered in greater detail in the next section (see also Chapter 6 for scope usage in signal tracing).

4-6 TESTING AND TROUBLESHOOTING WITH A SCOPE

In initial troubleshooting and servicing procedures the VOM units are generally used to localize defects. When ordinary voltmeter and ohmmeter checks are inadequate to help localize the trouble, the scope facilitates the pinpointing process by displaying the faults, if any, existing in the signals at both the input and output sections of circuits. Often, from experience, the technician can ascertain (from symptoms present) whether matters can be expedited by using the scope initially.

In stereo equipment the scope is useful for comparing the performance characteristics of the two channels. Similarly, for quadraphonic sound the waveform amplitude and shape can be compared to check for any differences among the several channels. Music or speech signals vary considerably in waveshape and amplitude, hence a square wave or sinewave signal of constant amplitude should be used to avoid misleading interpretation of scope patterns.

The input signals can be obtained from a signal generator applied to the inputs of the amplifier system. A test tape can also be used, in which case the tape deck replaces the signal generator. For phono units, test records are also available that provide a variety of sounds to check the complete system for separation, distortion, noise levels, signal amplitudes, balance and so forth.

Since the tape deck or pre-amp system could be contributing to any troubles present, the output from such units should be observed on a scope to see that no distortion or amplitude differences are present for a particular channel. As shown in Fig. 4-18, the shielded cables are removed from the amplifiers and the scope input probes applied. A common ground must exist between the scope and the equipment under test.

If the tape deck or phono input checks out satisfactorily the shielded cables can be applied to the amplifiers for sound input, or a signal generator can be used as mentioned earlier. Now scope patterns are observed at the output lines to the speakers as shown. Amplitude differences between channels will occur, of course, if the volume controls are not set at the same ohmic value, or if the balance control is off center. When controls are properly adjusted, the sinewave or square wave obtained at the output of each channel should be identical to the other signals observed. If one channel displays a weak signal or a distorted output compared to the others, localization of the defect must be undertaken by VOM tests or signal tracing as detailed in Chapter 6.

In the observation of high-frequency signals some distortion can be introduced by the bandpass limitations of the vertical amplifiers of the scope, as mentioned earlier, or by use of unshielded input cables to the scope. Reference should be made to the manual accompanying the

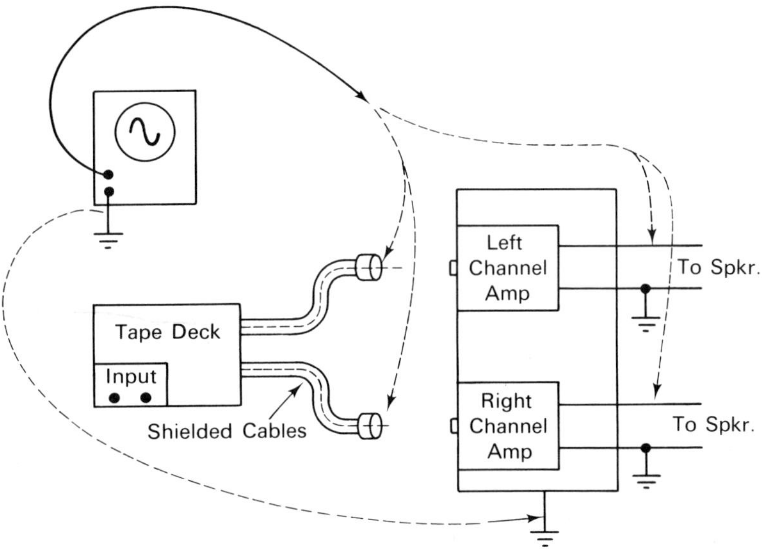

Figure 4-18 Testing stereo tape system.

scope for details regarding use of the proper scope probe to minimize stray capacitance losses and interference picked up by the scope probe because of unsatisfactory shielding.

Another consideration in the interpretation of scope patterns is the variation that could be the result of the settings of a scope during signal observation. From experience the technician soon recognizes width and amplitude variations caused by scope control settings, and actual signal distortion caused by defects in the circuits under test. These factors can be illustrated by a comparison of a typical schematic with waveforms drawn in and those observed on the scope. The circuits shown in Fig. 4-19 will serve as a representative example.

Note the pulse waveform at the collector of the pulse shaper transistor. Negative pulses are shown with a peak-to-peak amplitude designated as 20 V. When a scope is applied between this terminal and ground the pattern displayed may be as shown in Fig. 4-20. From a casual glance one might assume that something must be seriously wrong with the circuit since the waveforms apparently do not coincide. Actually, however, if the waveform displayed in Fig. 4-20 has the necessary 20 V peak to peak (PP), circuit function is normal. The pattern in Fig. 4-20 is inverted and can be changed into a negative display by flipping the polarity switch of the scope. Also, if two separate input leads are used instead of a shielded cable, a reversal of lead positions to the circuit will cause a polarity change.

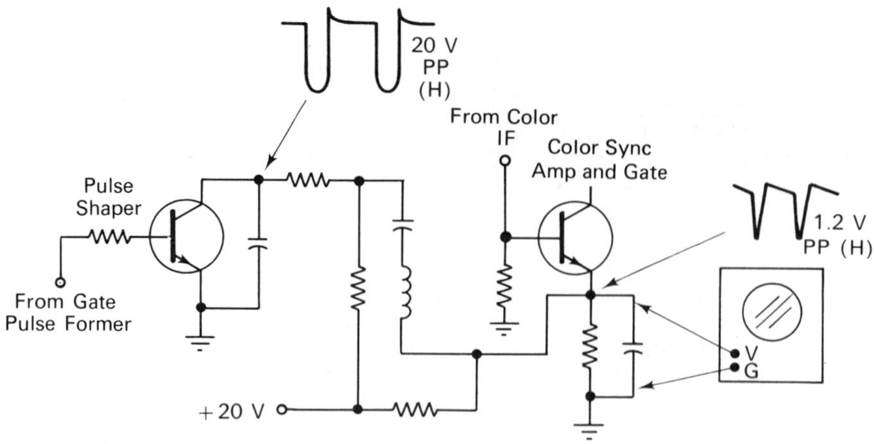

Figure 4-19 Color circuits scope patterns.

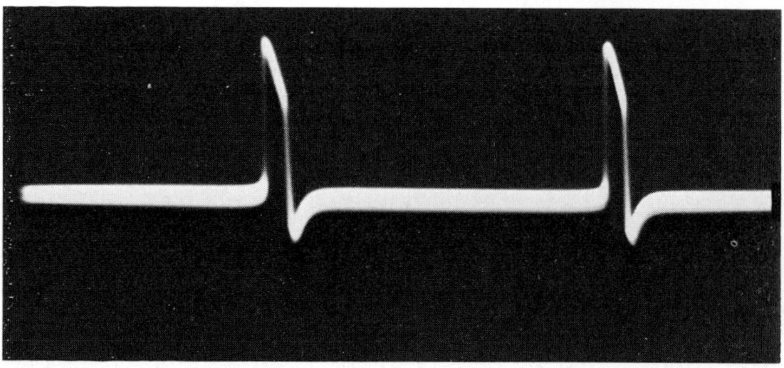

Figure 4-20 Pulse waveform.

The pulses in Fig. 4-20 are separated more than are those in Fig. 4-19. This is of no significance, however, since they can be brought into closer proximity by the horizontal amplifier control of the scope. The sharper points on the peaks of the pattern of Fig. 4-20 may be caused by a slight overshoot in the scope, or may actually represent the wave shape; the one shown in Fig. 4-19 may not be drawn too accurately because waveform amplitude and not shape is the important factor here.

Similarly, the pattern drawn at the emitter of the color sync amp and gate in Fig. 4-19 may appear as shown in Fig. 4-21, where slight differences occur because of control settings and scope characteristics. Again, if the observed pattern has a peak-to-peak value of 1.2 V circuit function probably is normal. (Phase differences in certain color circuits are important, however, and critical adjustments are sometimes required.

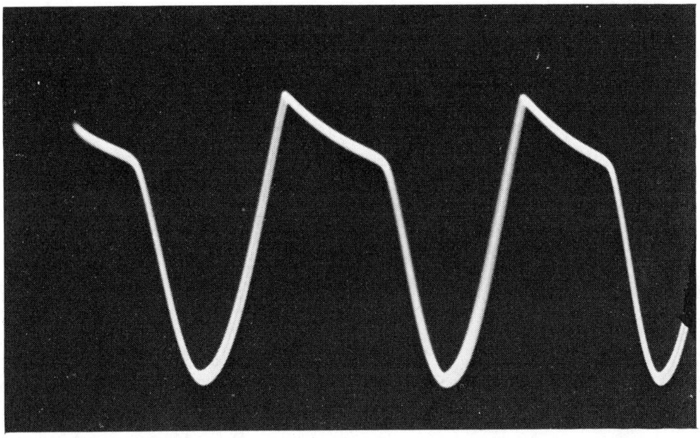

Figure 4-21 Sync waveform.

Additional information is given in Chapter 10. Other oscilloscope references will be found in Chapters 6, 9, 10 and 12.)

Another example is shown in Fig. 4-22 in which a phase detector schematic is shown. This circuit compares sync pulse timing from the transmitted TV signal with that generated by the horizontal oscillator to assure good lock in of the sweep. Note the 3 V PP signal at the upper anode of the phase detector. Here a pure sawtooth signal is shown that has a gradual incline and an abrupt decline. The feedback signal at the lower anode has a rating of 45 V PP and has a slight curvature in both

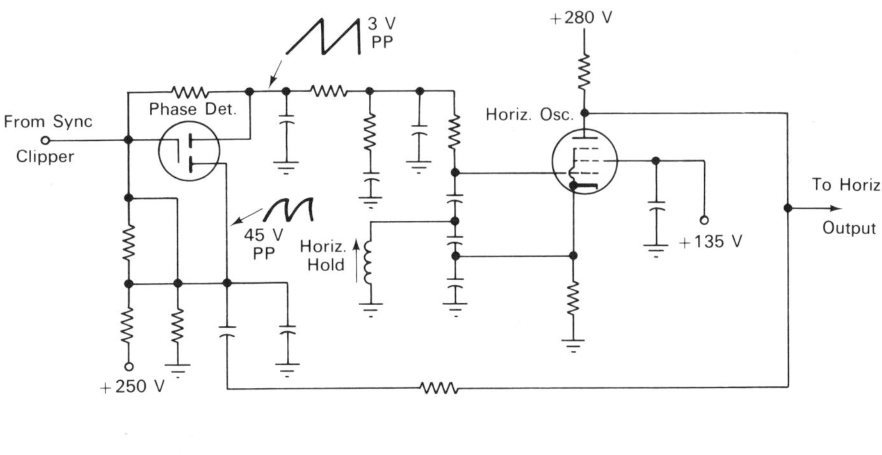

Figure 4-22 Phase detector waveforms.

the incline and decline. If, however, the observed waveform is as shown in Fig. 4-23, it indicates distortion because of the gradual decline (loss of high-frequency signal components). Hence, horizontal instability would occur and adjustments of horizontal hold would be critical. Thus, the phase detector tube (or transistor, if used) and associated components must be checked to localize the defective units. Tests would also include reading the input voltages (as well as the applied voltages to the horizontal oscillator and that appearing at the anode).

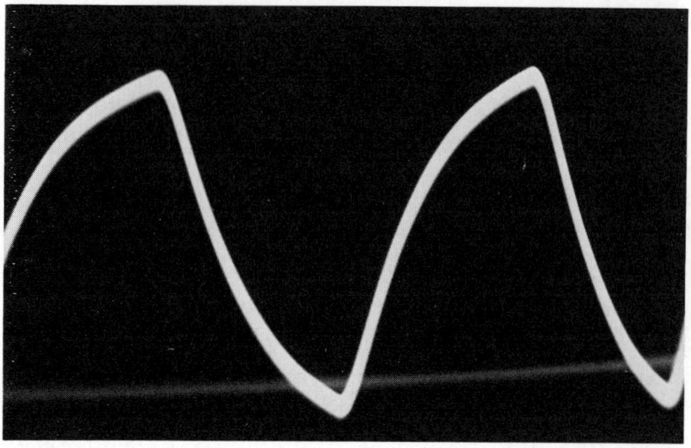

Figure 4-23 Phase detector signal distortion.

5

tests of C, L, X, and Z

5-1 C, L, X, AND Z FACTORS

The unit values in microfarads (μF) of the capacitance (C) of a capacitor is usually designated on the schematic for the electronic circuit in which it is used. Similarly, the unit value in henrys (H) of the inductance (L) of a coil is also generally specified. (Fractional values, of course, will also be found, such as mH, μH, etc.) Such information is provided so that proper replacements can be obtained when defects occur.

Inductor and capacitor defects usually appear as leakage, open-circuit conditions, shorts or partial shorts. A change of inductance or capacitance (without other contributing defects) is rare. Hence, tests and checks of such units consist of ohmmeter readings or leakage tests with inductor or capacitor checkers. In design and laboratory procedures, inductance and capacitance value readings are undertaken more often than during servicing or troubleshooting procedures.

Reactance (symbol X) has for its unit value the ohm (Ω) the same as with a resistor because the reactance is the opposition offered to alternating current by either an inductor or capacitor. Because of the phase displacement caused between current and voltage, however, the power is altered as discussed in Sec. 5-3. Reactance in an inductance (L) is specified as X_L while that of a capacitor is X_C. Impedance (Z) is formed by the combination of R with either C or L.

Reactance and impedance for ac, as with dc, follows Ohm's law (see Appendix A). As with L and C, it is rare that the need for X and Z measurements arise in testing and troubleshooting procedures except in design laboratories. The Z factor is encountered more often because of matching procedures necessary with transmission lines and antennas, output circuits and load systems. Laboratory instruments are available for direct measurement of L and C, utilizing the *bridge* circuit for comparing unknown values with known standards of measurement as discussed in Sec. 5-2.

5-2 BRIDGE-CIRCUIT METERS

A symmetrical arrangement of resistors, capacitors and inductors, or combinations of such units, can be formed for measurement of unit values of resistance, inductance, capacitance, and impedance. The common circuitry is the bridge, also referred to as the *Wheatstone bridge* after Charles Wheatstone, the English physicist who first disclosed the value of such a balanced circuit arrangement.

The basic dc bridge circuit for finding the value of unknown resistors is shown in Fig. 5-1A. Resistor R_s is a standard resistor of known

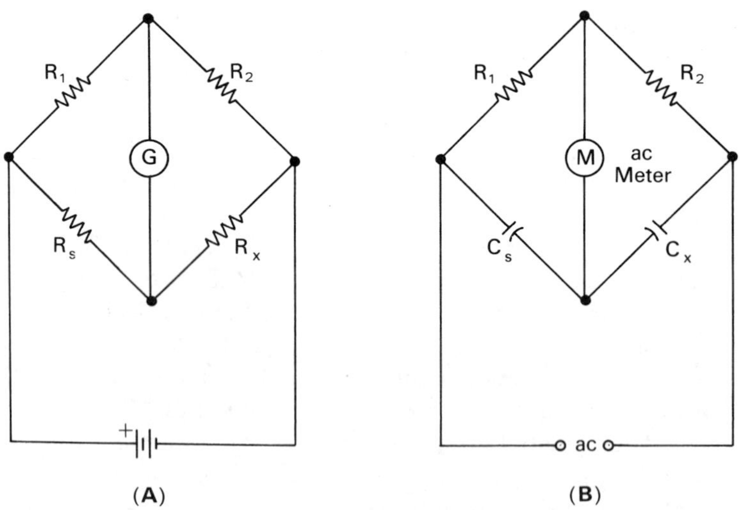

Figure 5-1 R and C bridges.

value that can be a variable type. A calibrated dial may be used to vary the resistance and read the ohmic value directly. Resistor R_x is the unknown resistor to be measured. When the bridge is balanced the galvanometer (G) reads zero (see Sec. 2-3). A microammeter can be used instead of the galvanometer, though a zero-center meter is preferred.

When the bridge is balanced, the resistor relationships creating the balance may be set down in equation form as follows:

$$R_x = R_s \frac{R_2}{R_1} \tag{5-1}$$

A capacitance bridge is shown in Fig. 5-1B. Here, the capacitors of the known and unknown value are designated as C_s and C_x. Now ac is applied as shown (the meter must be an ac type). For the capacitance bridge the equation for finding C_x has the R_1 and R_2 designations reversed because the reactance of a capacitor *decreases* for higher values of capacitance:

$$C_x = C_s \frac{R_1}{R_2} \tag{5-2}$$

An inductance bridge can also be formed by replacing the C_s and C_x capacitors with inductors. With inductors, however, the reactance increases for higher values of inductance, hence the equation resembles the one shown for resistance:

$$L_x = L_s \frac{R_2}{R_1} \tag{5-3}$$

In commercial bridge-type meters additional components are included so that direct-reading dials show the value of the unit under test. The capacitor meter, for instance, would include the transformer for furnishing the necessary ac potentials, and a switching arrangement for selection of the appropriate C_s value, as shown in Fig. 5-2. Resistors R_1 and R_2 are combined in a single potentiometer as shown, and the shaft of this variable resistor connects to the meter dial. A tuning-eye tube such as the 6E5 or a meter is used to indicate bridge balance. Depending on the value of the standard capacitances used, meter readings range from a few picofarads to several thousand microfarads in overlapping ranges.

The controls brought out on the front panel include those shown in Fig. 5-3 in commercial types. The indicating eye shows bridge balance and the meter at the upper left is for reading leakage current of electrolytic capacitors in current ranges of 0.05, 0.5, 5 and 50 mA. The meter is

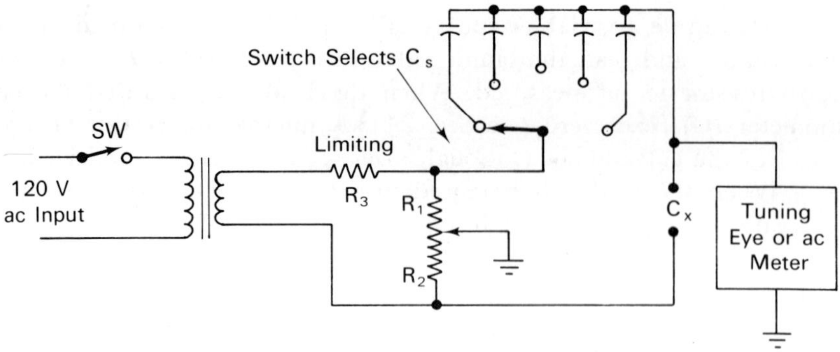

Figure 5-2 Basic capacitor meter.

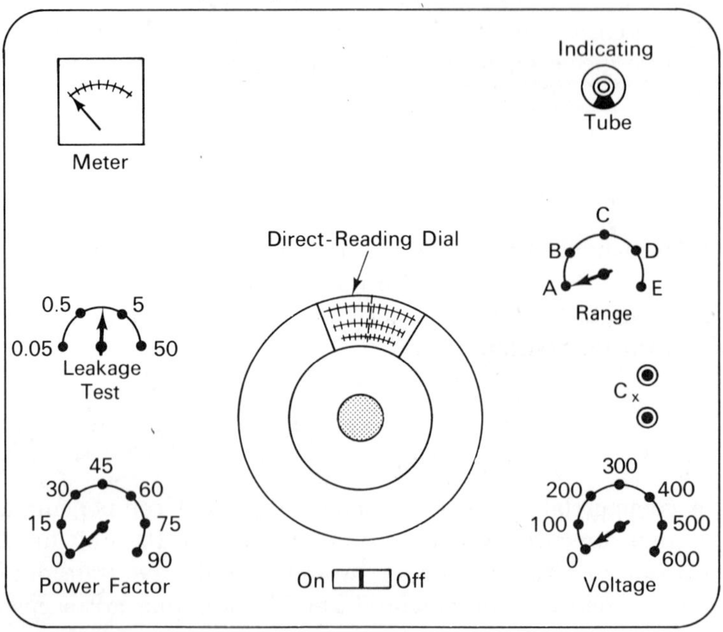

Figure 5-3 Capacitor meter panel controls.

also used for testing insulation resistance of paper, mica and ceramic capacitors.

The range of the capacitance read on the dial is selected by the range switch where *A* provides for the low-value range, and *E* the maximum-value range. The voltage switch selects the values needed for conducting leakage current tests and insulation checks.

The lower-left switch is used for measuring the power factor of the larger-value capacitors (such as the electrolytics). With a defective capacitor having some leakage, a resistive element is introduced that reduces the capacitor's effectiveness as a filter. Thus, a low power-factor rating is preferred and capacitors that check out in the high power-factor category are discarded during servicing and troubleshooting procedures.

For measuring the large-value low-working-voltage capacitors used in transistor circuitry, special low-voltage checks are available to avoid capacitor damage during checking procedures with the capacitance analyzers.

5-3 POWER FACTOR

When effective values of ac flow through a resistor, the amount of electric power consumed (in watts) is the same as it would be for dc (see Sec. 2-4). The simple equation $P = EI$ can be misleading, however, when some inductance or capacitance (or both) is present in the circuit. Inductors and capacitors affect the phase relationships between voltage and current and when the two are not in phase the amount of power consumed is reduced. The reduction is caused by the fact that the peak values of current and voltage no longer coincide.

With out-of-phase conditions the equation $P = EI$ solves for the *apparent power*. (The P, E, I designations are assumed to be the rms or effective values, since these are the most commonly measured by general test equipment.) The apparent power is *not* the *true* power when there is a phase difference between current and voltage. With no phase difference, the true power equals the apparent power. With a phase difference, the equation for true power in rms values is:

$$P = EI \cos \theta \tag{5-4}$$

Thus, when E and I are in phase, the angle is 0° and the cosine is 1; hence the apparent power equals the true. For a phase difference, however, the angle has a cosine less than 1, hence the apparent power is reduced (see Appendix F). If the effective values of E and I are 20 V and 5 A, for instance, the apparent power is $20 \times 5 = 100$ W. If there is a 60° angle between current and voltage, however, the cosine is 0.5, and consequently the true power, using Eq. 5-4, would be:

$$20 \times 5 \times 0.5 = 50 \text{ W}$$

Meters are available for reading the power factor (cosine) directly. The meter is used in series between the ac line and the load and the front panel may appear as shown in Fig. 5-4. Often the dial pointer does not have the usual spring return for bringing it back to zero (or the center position of 1 for this meter). When a particular phase angle between current and voltage is read on this meter, the pointer remains at one reading until another measurement is taken. If a new phase angle is read, the pointer repositions itself accordingly.

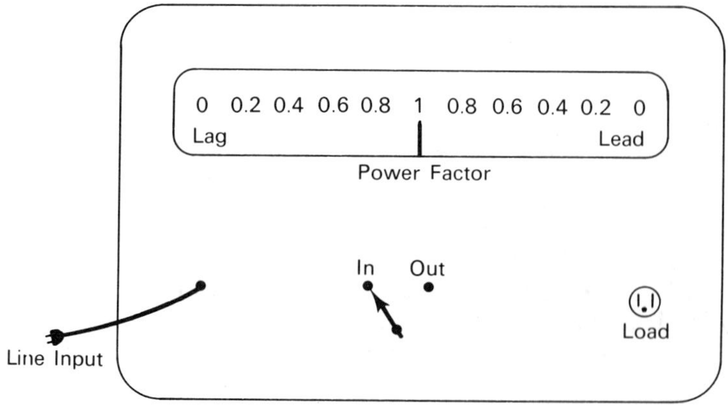

Figure 5-4 Power factor meter.

A power factor meter is generally not needed in testing and troubleshooting electronic equipment. In industrial electric installations, however, it is often necessary to correct for power factors less than 1 to reduce excessive currents. With industrial loads such as motors and other units not purely resistive, inductive effects are produced and cause phase angle changes. Additional currents circulate, though these currents do not cause power consumption in inductors or capacitors (a pure reactance does not consume power). The additional current circulates through the reactance and other devices and is returned to the generator, hence the additional currents must be furnished by the power companies.

The watt-hour meter commonly used in homes to measure power consumption indicates only the true power consumed and does not indicate the greater circulating current for low-power factors. Hence, in industry, volt-ampere-hour meters may be installed or penalty rates imposed in proportion to the power factor that is less than 1. To compensate for current-lagging conditions, large capacitors are installed by consumers to correct the low power factor. This brings the current back to normal and prevents excessive voltage drops across resistors and other load systems, which can cause generator and transformer overload conditions.

5-4 FINDING AND CORRECTING C- AND L-RELATED TROUBLES

An ohmmeter set on the megohm range is also useful for testing electrolytic capacitors as well as the larger-value paper and ceramic types. Electrolytics will have a low-ohm reading in one direction and a much higher reading when the ohmmeter probes are reversed across the capacitor leads. Electrolytics are marked for polarity and must be correctly wired into the circuit to prevent damage. Paper and ceramic capacitors, however, have an infinitely high resistance in both the forward and reverse directions. When a narrow band is imprinted at one end it denotes connection to the outer foil of the capacitor, which can serve as a shield for the remainder of the capacitor windings. Thus, the banded end is usually placed at ground potential. For electrolytics having more than one unit in a case or can, semicircles, circles, squares and triangles are often used for identification, as shown in Fig. 5-5.

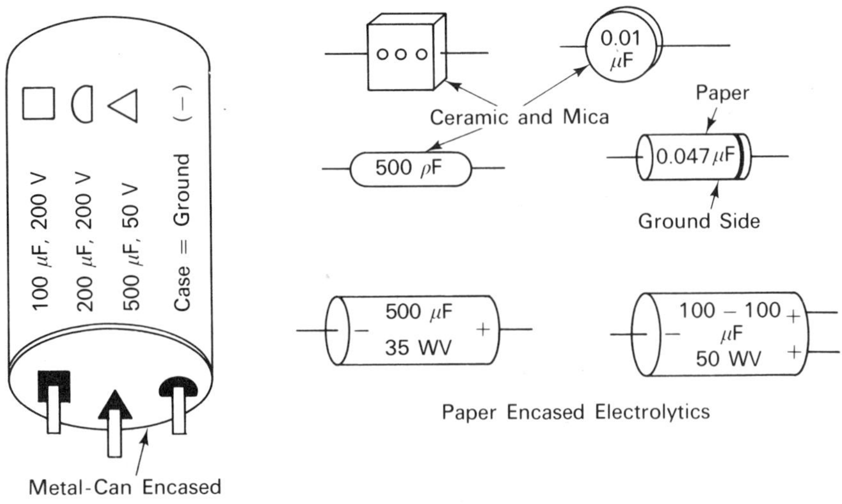

Figure 5-5 Representative fixed-value capacitors.

Ceramic, mica and paper capacitors usually do not have voltages marked on them, though many have ratings over 500 V hence present no problem in most usages. Capacitance may be marked on them, or color codings used as explained and illustrated in Appendix H. Paper-encased electrolytics, as with the can types, have capacitance and voltage markings, with the WV (working voltage) usually designated. Reference should be made to Sec. 1-7 for additional details on capacitor troubles. As also explained in Sec. 1-7, the voltage ratings must be considered when replacement is necessary, to ascertain that the new capacitor has as much (or more) WV as the old one.

Capacitance value should equal the replaced unit, of course, though in some instances a higher-value capacitance could be used if an exact value is not available. Some circuits, however, are critical with respect to exact values (including sweep oscillators in television receivers, oscillators in tuners, and phase-detector circuitry). Even with coupling capacitors, a larger value than the original one may mean greater physical bulk, which presents problems of replacement and may cause capacitance leakage of high-frequency signals to nearby components or the chassis.

In the absence of a capacitor checker, a quick test can be made for leakage with the ohmmeter, as mentioned earlier. The best checks are made with a very high ohmic reading, above 1 MΩ. For capacitors of 0.001 μF or more, a 1 to 5 MΩ range may be sufficient, though a range extending above 10 MΩ is preferable to assure that small leakages are read. For some coupling capacitors (and frequency-determining units in oscillators) a slight leakage is sufficient to upset circuit function. Thus, one side of the suspected capacitor is disconnected and a ohmmeter reading taken. The meter needle should show a low resistance momentarily and climb rapidly to a high ohmic value if the capacitor has no appreciable leakage. If the meter needle fails to climb substantially above 1 MΩ, the capacitor should be replaced. Values below 0.001 μF are difficult to check with the ohmmeter unless they are shorted or partially shorted. For the smaller values, a direct replacement should be tried if no capacitor checker is available.

Schematics may show the inductance value for coils or, if the value is very low, may just indicate the ohmic value. Thus, testing is simplified since an ohmmeter check will suffice and an inductance bridge is not needed. A typical example of *C* and *L* designations is shown in Fig. 5-6 for the circuit between the last video amplifier and the picture tube. At the output of the video amplifier a transformer is used for sound IF take off. Thus, the primary of the transformer is shunted by a 47 pF capacitor to form a resonant circuit for the 4.5 MHz frequency of the sound IF. Note the critical value of this shunt capacitor (+ 5%) to stay within the tuning range of the signal. The primary inductance is not given, but the resistance is shown as 14 Ω. Thus, any deviation from this value calls for a check of the 47 pF capacitor for leakage, or for transformer replacement.

For the secondary a resistance of 2 Ω is shown. Thus, ohmic checks can be made of both primary and secondary if troubles are indicated. If the sound output is weak the tunable core should be adjusted to peak sound output. If sound levels are still too low, the sound IF,

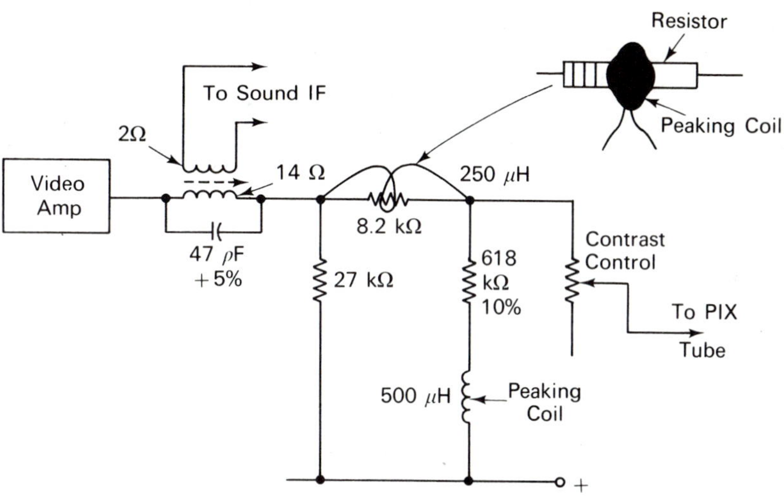

Figure 5-6 C and L factors in video amplifiers.

detector and sound output circuits should be tested before attempting transformer replacement. Low volume could be caused by weak tubes, defective transistors, improper voltages or defective components such as capacitors and resistors. If all these check out, then the sound output IF transformer should be tested. If the picture is also weak, the troubles may exist prior to the video output circuit (see Chapter 10).

In video amplifiers small inductors called *peaking coils* are used to improve fine detail in pictures and thus present a sharper picture. Such peaking coils may be used in series with the signal, in shunt across the circuit, or both, as shown in Fig. 5-6. The series peaking coil tends to isolate input and output signal shunts, while the shunt peaking coil (in conjuction with circuit capacitance) forms a parallel-resonant circuit of high impedance, thus minimizing signal-shunting effects.

The series peaking coil is often wound around a resistor. With resistance the sharp-resonant effects are broadened to prevent overpeaking. Thus, for the circuit shown, a resistor of 8.2 kΩ shunts a 550 μH peaking coil. Defects in either the resistor or coil would necessitate a complete replacement, since the two are often assembled as a unit, preventing separation of the coil or resistor. Replacement requires an exact type to avoid loss of fine detail in the picture.

Lack of crispness in the picture may also be caused by a defective shunt peaking coil. For Fig. 5-6 a 500 μH value must be replaced if found defective. If the shunt peaking coil opens for Fig. 5-6 it will upset contrast because of the removal of the positive voltage applied to the top of the contrast control. If the series peaking coil opens, its low resistance will no longer provide the necessary low-ohm coupling to the picture tube. Instead, the 8.2 kΩ resistor tends to limit signal transfer, hence contrast (and fine detail) will decrease. If the resistor opens but the series peaking coil has continuity, excessive fine detail will result in the formation of thin repeat lines at the edges of sharply-defined objects, as detailed in Fig. 5-7.

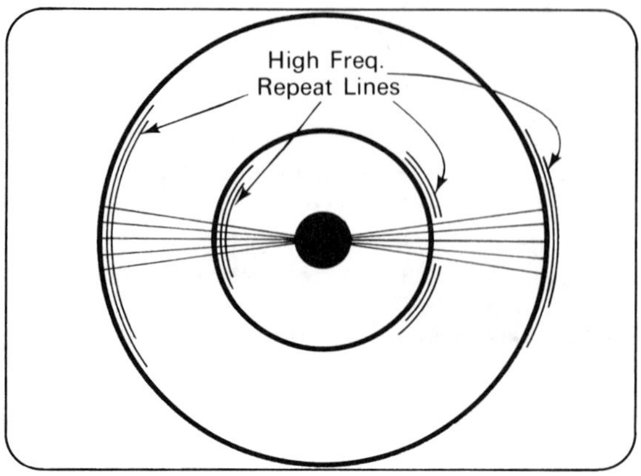

Figure 5-7 Effects of over peaking.

When replacing peaking coils they should be spaced well away from the chassis or from other circuit components. If this is not done, coupling effects are set up and the functions of the peaking coils are diminished. With stray capacitance set up between peaking coils and chassis, for instance, the decreasing reactance to higher frequencies shunts signals that are needed to provide for fine detail in the picture.

Peaking-coil circuitry is also found in the vertical amplifier stages of oscilloscopes capable of handling high-frequency signals. Some circuits of this type have also been used in broad-band high-fidelity amplifiers.

In television servicing procedures the lack of fine detail usually calls for an initial check of the focus control setting. This control is adjusted clockwise and counterclockwise until the horizontal trace lines are clearly defined upon close inspection of the screen. If fine focus can't be obtained by complete rotation of the focus control, the focus circuitry must be tested to see if focus voltages have dropped below normal (see also Chapters 7 and 10).

Another factor that contributes to loss of fine detail is a mismatch between the transmission line and the input to the receiver as detailed in Sec. 5-7.

5-5 VARACTOR TUNING DIODES

Because diodes have unidirectional characteristics, a simple ohmmeter check is often sufficient to ascertain whether the unit will pass current in one direction but not in the other. Also, the ohmmeter check will establish whether the diode forms an open circuit because it is defective (see also Chapter 8). Often, however, a defective diode is pinpointed because its malfunction causes symptoms pointing directly to it.

A representative example is the varactor diode, which simulates the capacitor because it exhibits a capacitance directly proportional to the applied voltage. The varactor *(variable reactance)* is sometimes referred to as a varicap *(variable capacitance)* with identical functions, since a change of capacitance automatically alters the capacitive reactance. When such a diode is connected across a resonant circuit or used as a shunting capacitance for an inductance, the resonant frequency can be changed by increasing or decreasing the applied dc. Thus, such a diode permits the elimination of variable capacitors for tuning purposes and simplifies the mechanical aspects of tuning while also reducing the size of tuners by using push-buttons, variable resistors or rotary switches for tuning purposes. The basic circuitry is shown in Fig. 5-8.

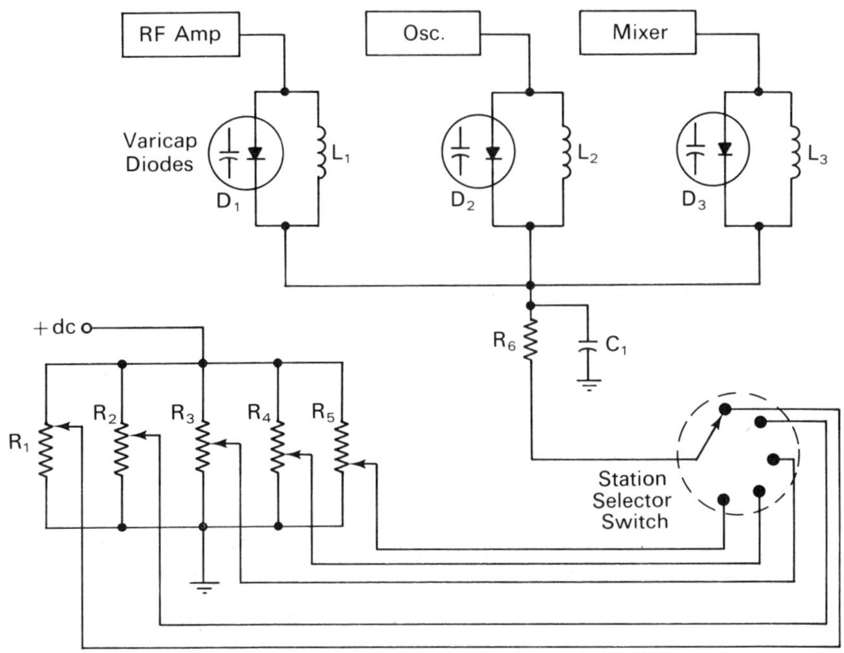

Figure 5-8 Varicap tuning.

The varactors shunt the tuning inductances as shown, and the voltage change is initiated by rotating the station-selector switch. Thus, appropriate resistors are selected for changing stations. Each variable resistor is adjusted to produce the voltage necessary to tune to a specific station (R_1 to R_5 in Fig. 5-8.) Resistor R_6 limits the applied voltage to that required, and C_1 places the bottom of the resonant circuits at signal ground by virtue of its very low reactance for signal frequencies.

If such a system fails to receive only one or two stations, the adjustments of the variable resistors must be checked. Tuning (of the resistors) may be off, or a defect may have developed in one of the potentiometers. Also, the voltage applied to the resistors should be checked against the schematic for the receiver to make certain that low voltages are not contributing to the trouble. If resistors R_1 through R_5 check out satisfactorily, resistor R_6 and the shunting capacitor C_1 should be checked. If R_6 is open, however, the station selector switch becomes ineffective for all stations. If C_1 develops a partial short, the voltages applied to the resonant circuits will be incorrect. Some stations may be tuned in, but they will not be the ones that were selected by the R_1 to R_5 settings. If all components below the varicaps check out satisfactorily, the diodes can be checked for proper forward-to-back resistance changes. Scope checks can also be made to see that the RF signal is entering the RF amplifier.

When replacing the varactors, a matching unit must be used because units exhibit different capacitances for certain voltages and have tuning ranges that differ. Once the new diode has been installed it will be necessary to retune the voltage-selector potentiometer to center-tune the desired station.

5-6 REACTANCE CIRCUITRY TROUBLES

Circuits having reactance characteristics can also be formed by transistors (or vacuum tubes) which also provide for a variable capacitance by changing the voltage applied to the input. A typical *pnp* transistor version is shown in Fig. 5-9A. The functional components making up the reactance characteristic are capacitor C_1 and resistor R_1. The capacitance of C_1 is selected so its reactance is about ten times the ohmic value of R_1. The collector and emitter (via the signal ground established by C_2) shunt the inductor of the circuit where frequency is to be controlled. In color television receivers such a reactance circuit is used to stabilize the 3.58 MHz oscillator and to lock it in with the incoming 3.58 MHz synchronizing signal (see Chapter 10).

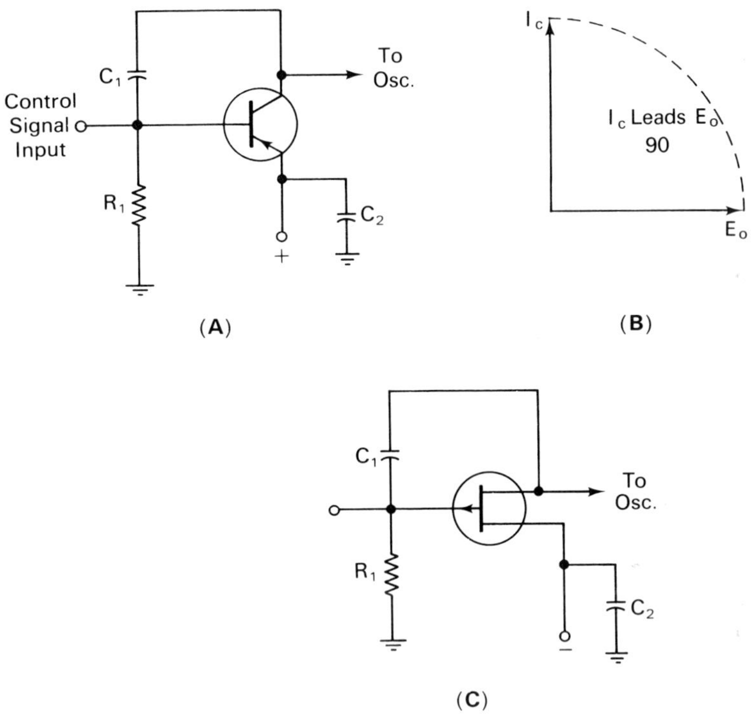

Figure 5-9 Reactance circuitry.

Since C_1 has such a high reactance, the circuit made up of C_1 and R_1 is primarily capacitive and current I_c leads the signal voltage E_o of the oscillator, as shown in Fig. 5-9B. Since the signal applied to the base of the transistor is obtained from across R_1 only, the voltage at the base will be in phase with the C_1-R_1 network current, because there is no phase shift across a pure resistance. Because the signal current through the transistor is in phase with the voltage at the base input, the collector current leads the oscillator signal by 90° as shown in Fig. 5-9B. Since a capacitor produces a current lead, the circuit at in Fig. 5-9A thus exhibits capacitance characteristics and hence capacitive reactance. (A vacuum tube was used in early receivers in place of the transistor, with identical formation of reactive characteristics.)

Figure 5-9C shows a field-effect transistor version of the reactance circuit. A P-channel unit is used to obtain the same type reactance as the transistor circuit shown in Fig. 5-9A. An N-channel FET could be used, of course, though bias polarities would be different. Similarly, an *npn* transistor could be used instead of the *pnp* shown.

When such reactance circuits are used with oscillators, manual tuning adjustments are provided so that the correction range of the reactance circuit extends over the frequency at which the oscillator is to be maintained. As shown in Chapter 10, phase detector circuits sample the incoming 3.58 MHz signal sent with the color signals and compare it to the frequency of the signals produced by the oscillator. Differences are corrected by the reactance circuits. Thus, an inoperative or intermittent reactance circuit can cause color contamination on the screen and false color rendition. Synchronization is critical, since even slight phase difference produce false color rendition.

When troubles occur, the tuning controls must be tested initially and if these are satisfactory the reactance circuit transistor and components must be checked individually. Additional data are given in Chapter 10.

5-7 IMPEDANCE MATCHING FACTORS

Transformers are often used in the output stages of amplifiers for isolation of dc components from the circuit to the output units (such as speakers, recording heads, etc.) and for impedance-matching purposes. As shown in Fig. 5-10 A and B, a step-down transformer is used to match the low impedance of the speaker's voice coil (between 4 and 16 Ω) and the relatively much higher circuit impedance at the plate circuit of the tube or the collector circuit of the transistor. Actually, however, the voice coil is not matched to the tube's plate resistance or the transistor collector-emitter impedance, but instead to an ohmic value designated as the recommended load resistance (R_L). This R_L value is that which provides the best results in terms of low harmonic distortion. For a maximum transfer of signal power, however, an impedance match is necessary. In most amplifiers some power output is sacrificed to obtain better performance. Hence, when transformer replacement is necessary, such impedance-matching factors must be considered and a proper replacement type obtained.

The turns ratio necessary for an impedance match is found by the following equation:

$$\text{turns ratio} = \sqrt{\frac{Z_1}{Z_2}} \tag{5-5}$$

For mathematical convenience, the higher impedance is considered as Z_1. Thus, if the recommended load resistance is 39,200 Ω and

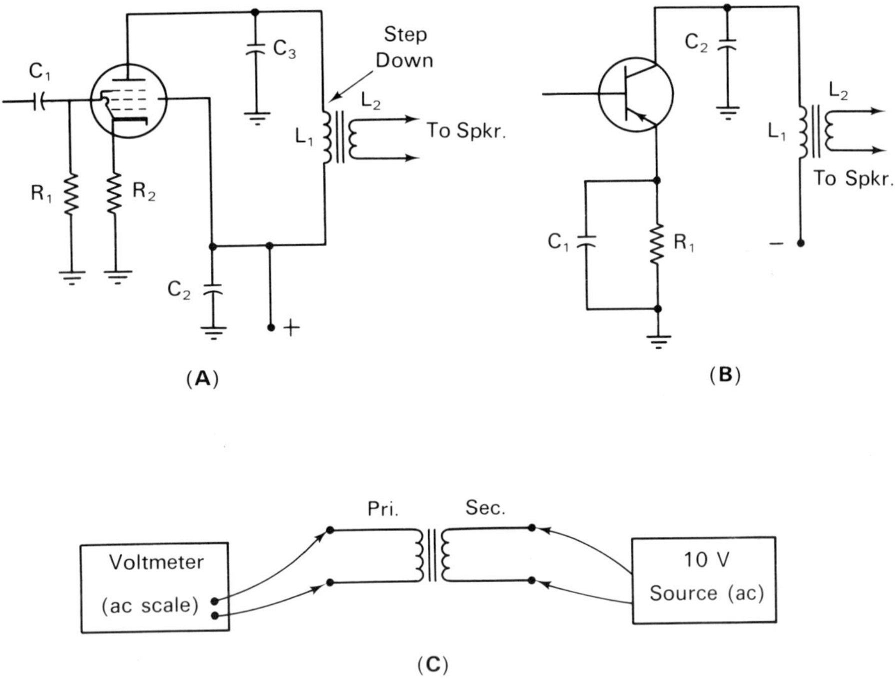

Figure 5-10 Z-matching factors.

the voice coil impedance is 8 ohms, the following turns ratio is necessary to match impedances:

$$\sqrt{\frac{Z_1}{Z_2}} = \sqrt{\frac{39{,}200}{8}} = \sqrt{4900} = 70 \text{ to } 1$$

Thus, the primary of the transformer must have 70 more turns on it than the secondary. Hence, the primary could contain 700 as against 10 for the secondary, or 7000 to 100 or 140 to 2.

In audio transformers it is possible for some of the primary turns to short out and upset the turns ratio and the impedance matching. For an impedance mismatch there will be a loss of output power, though this fact may not be apparent if the mismatch is slight. If, for instance, an audio amplifier with a potential of 50 W output has a slight mismatch develop, the maximum power available may drop to 40 or 35 W, but this fact would not be ascertained if the amplifier is operated at 10 or 20 W maximum output. For severe mismatching, however, the power loss is quite apparent and some harmonic distortion may also develop.

The turns ratio of a transformer may be determined by applying ac to one side and measuring the voltage obtained from the other. Thus, if 10 V is applied to the secondary and 100 V is read across the primary, the turns ratio is 10 to 1. (The voltage could be applied to the primary, of course, and compared to that obtained across the secondary. Thus, if 10 V is applied to the primary and 0.1 V is obtained from the secondary, we again verify a 10 to 1 turns ratio.)

Proper impedance matching is of some importance in antenna installations for FM or TV receivers, particularly in weak signal areas. Where standard twin-lead is used, the impedance has been standardized at 300 Ω and consequently tuners are designed with a 300-Ω input. On occasion, however, an additional input is provided with an impedance of 75 Ω to match coaxial cables. The coaxial cable transmission lines have a braided metal outer conductor usually connected to ground, which shields the inner conductor. Such a lead-in line is less prone to pick up noise signals and has been used in high-noise areas. The coaxial cable, however, has more signal loss for a given number of feet than the twin lead, hence signal gain suffers, but in a strong signal area little difference may be noticed.

In theory a perfect match should be obtained by using an antenna and transmission line rated at 300 Ω each. Unfortunately, however, antenna impedances vary around the 300 Ω value and not all transmission lines are exactly 300 Ω. Similarly, the input Z to the tuner may vary somewhat, and may change considerably for higher-frequency station tuning. Thus, reliance must be placed in antennas and transmission lines from reputable manufacturers in hopes of achieving a reasonable match. If all items match, as in Fig. 5-11A, a maximum transfer of signal energy occurs from antenna to receiver input. (Maximum signal pick-up by the antenna, however, depends on proper orientation with respect to the station. If the antenna does not point to some stations weakened signals will prevail for those stations. If stations are distant and scattered in various directions an antenna rotor is highly recommended.)

If a mismatch occurs betwen antenna and line, as shown in Fig. 5-11B, there will be less transfer of signal from antenna to line in proportion to the mismatch. A similar condition prevails in Fig. 5-11C, where the mismatch between antenna and the coaxial cable decreases signal strength, even though a match exists at the receiver. If a mismatch occurs at the receiver (such as using a 75Ω coaxial cable with a 300-Ω input to the tuner) not only will signal strength suffer, but some line reflections will occur (called standing waves). This comes about because the mismatch at the receiver does not permit the tuner to accept all the signal energy delivered by the line. Consequently, some is reflected back to the antenna and in turn is again sent to the receiver. Some loss of picture

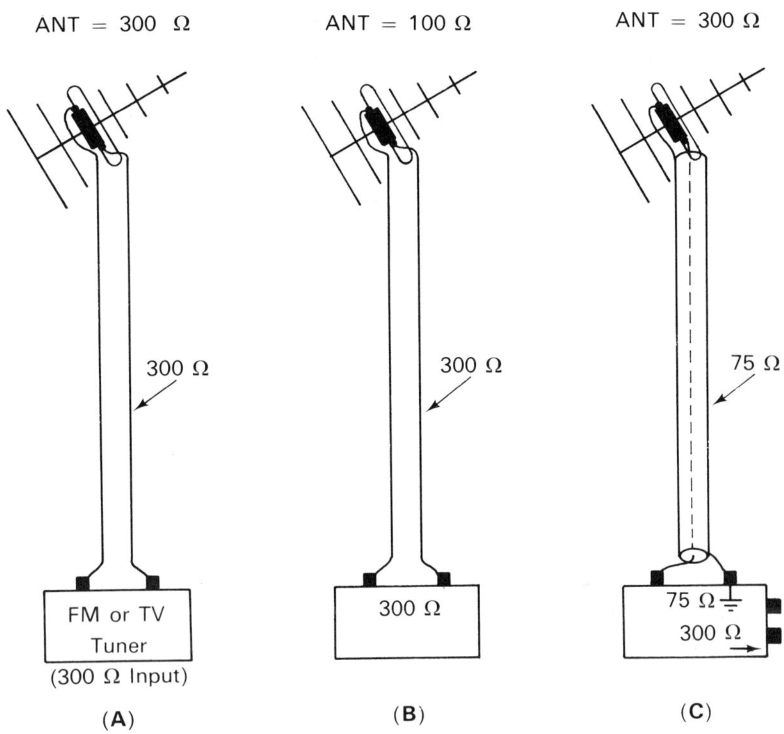

Figure 5-11 Antenna and line matching.

quality occurs and the result is that the picture appears slightly out of focus. Reflections cause repeat lines, but the length of travel from receiver to antenna is too short to cause a displacement to be visible on the screen. Instead, the added signal tends to add some fuzziness to the scene.

Mismatches can be corrected by using the special matching transformers available from suppliers. Such transformers match coaxial cables to 300-Ω tuner inputs and also match 300-Ω antennas to 75-Ω coaxial lines for improving reception.

After a transmission line and antenna system has been in operation for several years replacement should be considered. The elements tend to corrode antenna terminals and rods, and the transmission line may develop hairline cracks that absorb moisture and reduce signal strength. Broken antenna elements also reduce signal strength and may create interference problems because of the lower signal to noise ratio that develops.

When installing new antennas and transmission lines care must be exercised to avoid capacitance coupling to nearby tin roofs, metal rain pipes, conduits and other metal objects. When an antenna or transmission line is too near metal, the capacitance effect established between the antenna rods or the wires of the transmission line and the metal rain pipes, etc. causes a signal loss, which increases for higher-frequency stations. In weak signal areas this effect is more serious. Even in good signal areas losses occur that could impair good reception, particularly for FM and television UHF stations. Hence, stand-off insulators should be used to space the transmission line away from the metal mast and other nearby metal objects. Do not tape the transmission line directly to the antenna supporting mast or to metal rain pipe. (For added safety protection use UL approved lightning arrestors for both the lead in and feed lines to antenna rotators, antenna amplifiers, etc.)

5-8 ANTENNA COUPLERS AND TRAPS

Often it is desirable to use a single antenna for two television receivers, or for a television receiver and an FM set. It is not good practice to connect the antenna terminals of two or more receivers to a single antenna transmission line because of several undesirable conditions that occur. The oscillator (in the tuner) of each receiver may interfere with reception on the other set. Paralleling the two receiver's inputs halves the standard 300-Ω Z and increases losses. Even in a strong signal area some stations will suffer because of the single connection for multiple sets.

When more than one receiver of any type is to be used with a single antenna, commercially available couplers should be used that provide for dc isolation between the receivers and also maintain the proper impedance match. Typical of the internal construction of such couplers is the transformer arrangement shown in Fig. 5-12A. The lead-in from the antenna is applied to the primary of a VHF transformer arrangement as shown, with two or more receivers obtaining the signals from the secondary windings. Lower-cost couplers sometimes use resistors to provide for the proper Z matching. With resistors, however, some losses are introduced and in weak signal areas reception may be weakened for some stations. In selecting a coupler, a good quality type should be chosen so it has low insertion loss (very little decrease in signal strength when inserted between the lead-in and the receivers). Also, the design should be such that good interset isolation is maintained to prevent interaction and interference to one set from the other.

As shown in Fig. 5-12B, couplers are also available for connecting two television receivers plus an FM set to a single antenna. In noisy signal areas, special traps (as seen in Fig. 5-12C) are available from

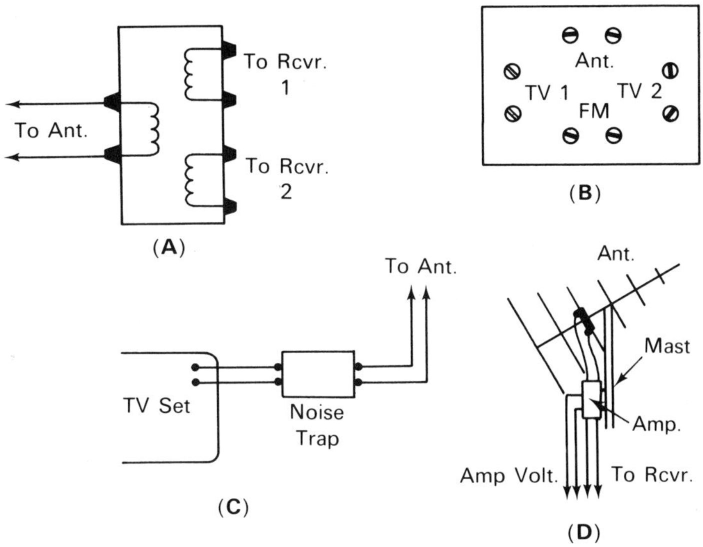

Figure 5-12 Antenna couplers, traps, and amplifiers.

electronic supply houses to filter out FM or TV station interference, ignition noises and interfering UHF station signals. For particularly weak signal areas booster amplifiers are available that can be mounted on the mast near the antenna. These increase signal gain from 10 to 15 dB. A power supply at the receiver is used with a special lead-in for furnishing power to the amplifier unit as shown in Fig. 5-12D.

6

signal tracing

6-1 SINGLE-SIGNAL GENERATOR

The term *single-signal generator* is sometimes used to distinguish between the basic generator of sinewaves or squarewaves of a selected frequency and the generators that sweep signals at a frequency rate and range needed for broad-band alignment, as covered in detail in Chapter 7. The basic signal generator is useful for signal-tracing purposes if it is capable of producing a signal having a frequency within the range handled by circuits under test. Thus, for checking AM radio IF circuits, the generator would have to generate signals of 455 MHz, while for testing audio circuits, a range of from 30 Hz to 30 kHz is sufficient. Thus, a versatile generator should be capable of generating both RF and audio signals, both of the sinewave and squarewave type. Also, the frequency should be variable and the output level voltage should be adjustable.

Signal tracing, as the term implies, is a procedure for ascertaining whether a signal is present in a circuit, and tracing such a signal through interconnected circuitry from the input to the output of an electronic system. The signal generator, by providing for a stable single-frequency signal, permits easier scope displays, voltage readings and signal identification. Thus, even though some signal tracing can be performed by using transmitted signals for radios (or microphones for audio amplifiers), the signal generator is preferred for reliable test results.

The appearance of the various signal generators differs as with all test instruments, though certain basic controls are common to all. A typical front-panel array of controls is shown in Fig. 6-1. A range selector determines the frequency span of a specific dial scale, and a switch at the right selects either RF (radio-frequency) signals or AF (audio-frequency) signals. The switch at the upper left determines whether the generator's output will be sinewave signals or squarewaves. The output level is controlled by the potentiometer at the upper right, and the coaxial-cable output jack is located at the lower left as shown.

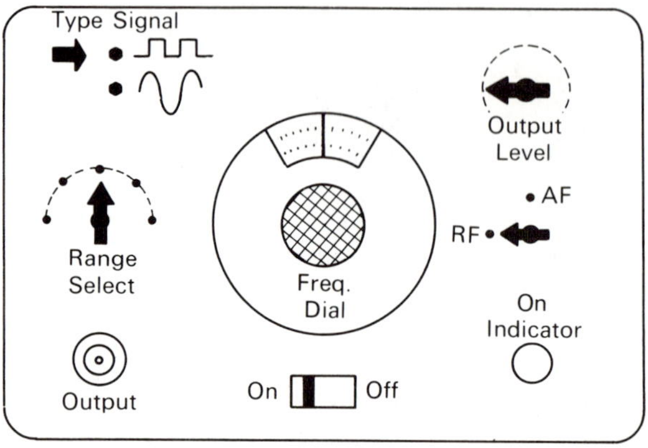

Figure 6-1 Signal generator controls.

Signal generators of this type have a frequency accuracy depending on how well they can be calibrated. Some have crystals incorporated into the circuitry for increasing frequency accuracy and stability, though such precise accuracy is not required for signal-tracing purposes. During alignment of IF stages, or the tracking of tuners, however, greater accuracy is needed, as detailed in Chapter 7. When accurate signal generators are used in conjunction with sweep generators, they are termed *marker* generators as described in the next chapter.

Most signal generators also have provisions for modulating the RF signal by an audio signal (such as a 400 Hz signal) so as to provide the type of AM signal that is transmitted. Thus, such a modulated signal can be injected into the tuner (or IF stages) and the detector will produce the 400 Hz tone output at the detector (and audio amplifiers) for signal-tracing purposes.

6-2 BASIC TRACING METHODS

Two basic methods for signal tracing amplifier stages are shown in Fig. 6-2. In Fig. 6-2A a signal generator is applied to the input of the first stage (a preamplifier, in this instance). If this is an audio amplifier, the input signal can be a 400 Hz tone, either sinewave or squarewave. The input could also be from a phono with a test record, or a tape player with a test tape. For signal tracing, an oscilloscope may be used as shown, as could an ac meter capable of sensing the signal present at each stage.

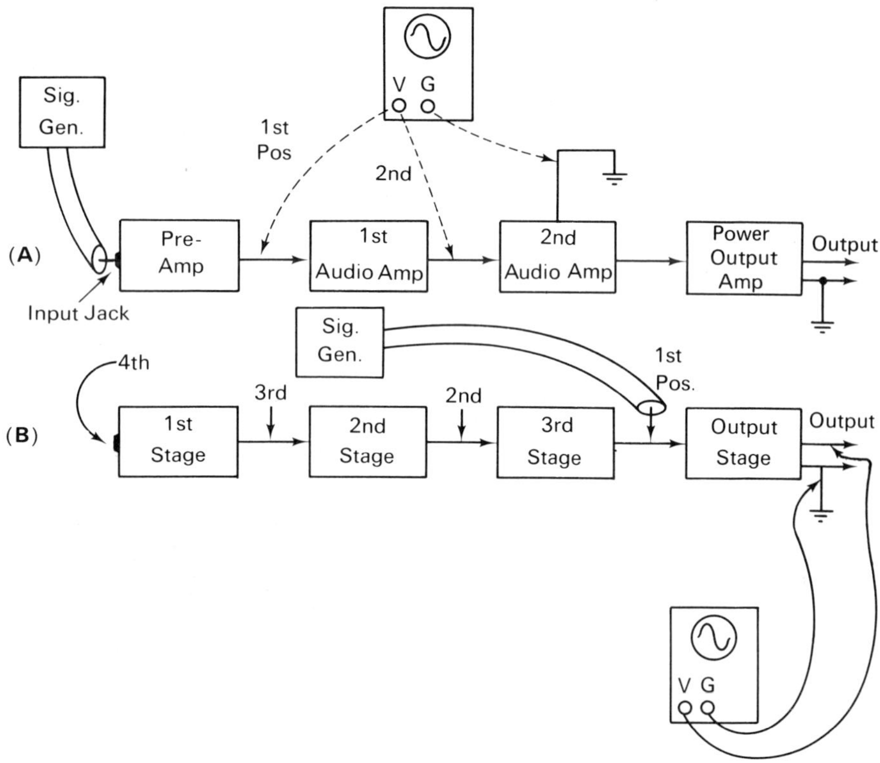

Figure 6-2 Two methods for signal tracing.

Thus, with a signal injected into the preamp, the scope input (vertical) is applied to the output of the preamp to check whether the injected signal arrives at the output (see Chapter 4). If no signal is received it indicates a dead preamp stage, and a VOM is then used to test for abnormally low or absent voltages, faulty components and shorts or partial shorts. If the signal arrives at the preamp output, however, the scope probe is moved to the output of the first audio amplifier and the absence or presence of the signal noted. If present, the scope is moved to the output from the second audio amplifier, and finally to the output of the last amplifier. Thus, the signal has been traced through each stage to localize the circuit that does not pass the signal (or passes it without proper amplification).

In the foregoing procedure the signal generator remained stationary at the preamp input and the scope was moved progressively through the various stages. Another method consists of leaving the scope in one position at the output of the amplifier and progressively moving the signal generator, as shown in Fig. 6-2B. The movement of the generator, however, starts from the last stage and moves left to localize the defective circuit. Thus, with the signal generator input applied to the output stage, the scope will display the pattern if this stage passes the signal. Once this is established we can apply the signal to the input of the third stage, and if this stage is satisfactory the scope will again display the injected signal. (Had the generator been applied to the third stage initially, the lack of signal output may have indicated a fault in either the third or output stage. Hence, the generator must progress one stage at a time to localize the defective one. Thus, if the signal is injected into the second stage and no signal is visible on the scope, this amplifier stage is obviously the one not passing the signal.)

A third method for tracing these circuits would be to move *both* the generator and scope. This procedure need not be done in any particular sequence, since it tests an individual circuit by applying the generator into the circuit under test and checking for a signal output with the scope from that same circuit. If a sequence is to be followed, this procedure can start at the final amp or at the first, as desired. Thus, a signal can be injected into the output stage and the scope attached as in Fig. 6-2B. Next, the signal can be applied to the input of the third stage, and the scope moved to the output of this third stage to check whether the signal is getting through. Next, the generator is applied to the input of the second stage, and the scope applied to its output, and so on.

6-3 SUPERHET SIGNAL INJECTION

The *superheterodyne* receiver principle is used in all public-entertainment receivers, including AM radios, FM radios and television receivers. As shown in Fig. 6-3, the tuner involves a mixing process between the incoming signal and that generated by an oscillator. The RF amplifier is sometimes omitted in the lower-cost receivers. The mixing process is a variable one, that is, as each station is tuned in, the oscillator's

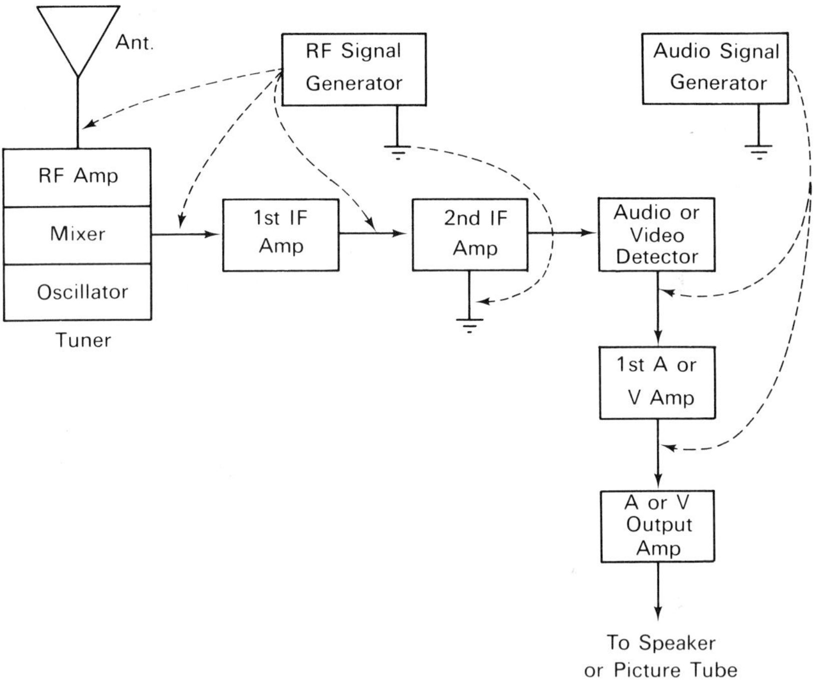

Figure 6-3 Superheterodyne signal applications.

frequency is changed so the mixing process always produces an intermediate frequency (IF) that remains the same for all stations received by the set. If this is an AM radio, 455 kHz is the standard generally used. For FM receivers 10.7 MHz is standard, and for television the video IF is usually 45.75 MHz and the accompanying sound carrier IF is 41.25 MHz. In a television receiver a second mixing process occurs at the detector, where the picture IF and the sound IF mix again to produce a special sound IF of 4.5 MHz. Additional information on this is given in Chapter 10.

For signal tracing in superhet receivers, several factors must be understood. As shown in Fig. 6-3, the signal generator must be one capable of producing a signal in the RF range for application to the tuner or IF stages. In addition, the generator signal must be modulated for convenience in detecting it at the demodulator output (either video or audio detector). If signal tracing is undertaken for the stages following the detector, the generator is set to produce an audio signal as shown.

If the signal generator is applied to the tuner input (at the antenna terminals) it must produce an RF signal at the frequency to which the tuner selector is set. Thus, if the AM radio dial is tuned to 900 kHz, the signal generator must be adjusted to produce a 900 kHz signal, modulated by an audio tone such as 400 Hz. Now, a scope placed at the detector will display the 400 Hz audio signal if each stage is functioning.

For signal injection in the IF stages, the generator must be set for 455 kHz and again modulated by a 400 Hz tone to provide for a scope display. An incoming signal from a broadcast station can be used instead of the signal generator, but this limits the procedure and requires scope observations at the output of the tuner and IF stages for pinpointing a troublesome tuner or IF amplifier. Since RF signals are involved, the scope must have a good frequency response and high sensitivity. On the other hand, with a generator, the scope can be left at the detector (to sense the low-frequency 400 Hz signal) and the generator moved back toward the tuner (stage by stage) until the defective circuit has been found. These factors also apply to FM radios and TV sets where it is possible to use the incoming signal for tracing purposes, but the signal generator and scope method facilitates the procedures.

The speaker (and picture tube) can also be used instead of the scope by injecting the signal into the last stage and progressively moving the generator back toward the tuner until the defective stage has been localized. With this procedure, however, one must make sure the speaker or picture tube is operating properly so as to function as the signal indicator during the tracing procedure. Proper connections to the various detector systems are covered in the succeeding sections of this chapter.

6-4 AM AND FM RADIO TRACING

The general form of the AM detector circuit is shown in Fig. 6-4. The last IF stage is coupled to the detector circuit by a transformer arrangement as shown, with the primary and secondary tuned to the 455 kHz IF signal. If off resonance, this stage can be tuned by the adjustable core of the transformer (or, in some instances, by a small-value variable capacitor as shown). The signal indicator unit is applied between the detector gain control and ground, as shown.

With a signal applied to either the tuner or IF a VOM will give a reading at the detector circuit. If the injected signal is not modulated by an audio tone, the VOM can be set on the dc scale to read the average unvarying dc produced by detecting the RF signal. If the injected RF signal is modulated, the VOM can be set on the ac scale to read the audio tone detected. An oscilloscope is applied to the same sections as shown for the VOM in Fig. 6-4 and adjusted to display the sinewave of

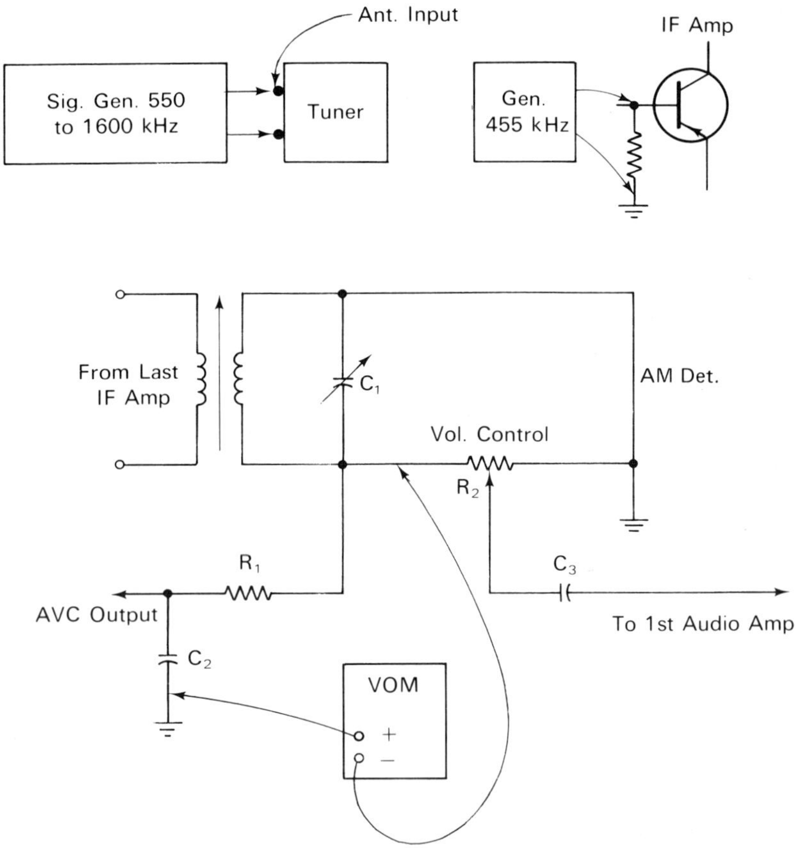

Figure 6-4 AM detector testing.

the 400 Hz signal used to modulate the RF signal being detected. If the RF signal is unmodulated, the scope must be adjusted to read dc levels.

Signal injection for FM radios is shown in Fig. 6-5 and connections to the FM demodulator are also shown. The circuit is a ratio detector and obtains the audio from the IF signal by detecting the changes in the voltage ratios developed across R_2 and R_3. Since the incoming IF signal is frequency modulated, the circuit compares the 10.7 MHz resonant center frequency with the signal frequency changes above and below the center IF frequency. The audio signal develops across R_3, which is in the form of a volume control. For the meter to detect an audio signal here, the injected signal from the generator must be *frequency-modulated* by the audio tone (400 Hz, or other audio frequency) (see also Chapter 9).

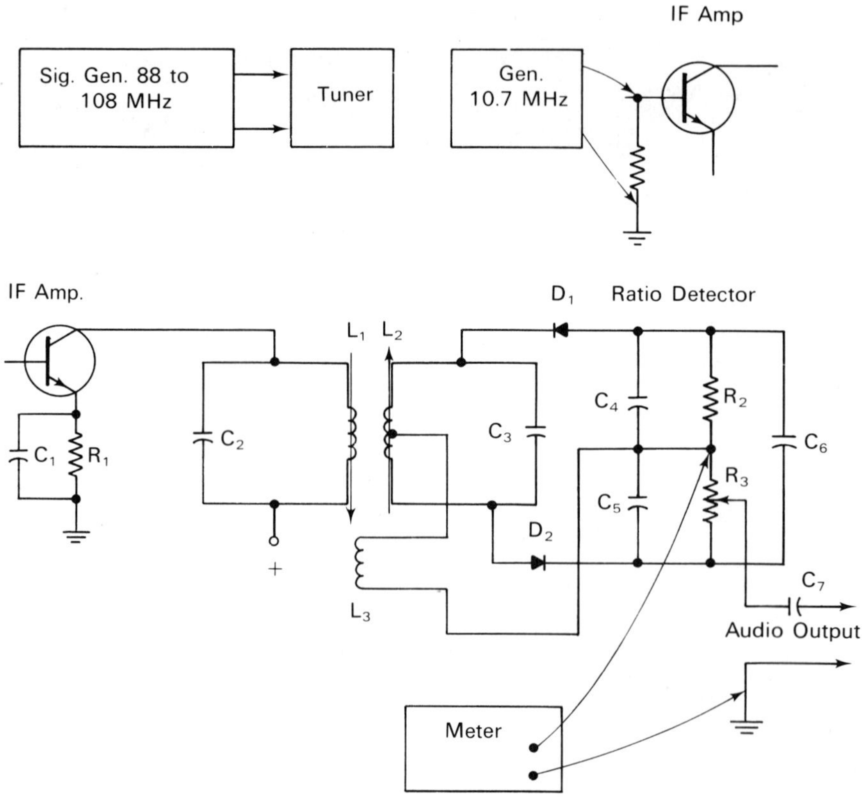

Figure 6-5 FM ratio detector circuitry.

Another type of FM detector is the *discriminator,* which has the diodes D_1 and D_2 wired in the same direction instead of the manner shown in Fig. 6-5. The signal is picked up across the volume control as for the ratio detector. Similarly, other detector systems, if used, present no special signal tracing problems. The VOM or scope is placed across the volume control and adjusted for the type signal being demodulated.

For multiplex detection of FM (stereo) plus block diagram of the stereo receiver see Chapter 9. For signal tracing the individual *audio* sections of stereo, follow the procedures outlined for single audio amplifiers described in Sec. 6-2.

6-5 TELEVISION SIGNAL TRACING

For signal tracing a television receiver from tuner to picture tube, a signal generator must be capable of producing signals in the frequency range of the VHF and UHF stations for tuner input. For IF input, the generator must be capable of supplying signals up to 50 MHz with provisions for modulation (see Sec. 6-3). Generators of the type described in Chapters 7 and 10, using broad-band principles, can also be used. As with AM and FM radios, the incoming signal from stations is also useful for special-purpose tests and tracing procedures.

A typical video detector is shown in Fig. 6-6, where the detector load resistor is R_1. It is from the top of this resistor to ground that the demodulated signal is obtained during signal tracing, testing, and alignment procedures. A solid-state diode is generally used for the video detector, though on occasion a triode transistor may be used as a power detector for greater signal output. Coils L_1 and L_2 are peaking coils for producing a sharper picture detail. Peaking coil factors and servicing will be found in Sec. 5-4.

As shown in Fig. 6-6, the vertical input probe of the scope picks up the detected signal for visual display. The frequency response of the scope need not be too high, since a demodulated signal is picked up here. The frequency of the signal depends on that used for modulating the RF signal injected into the RF or IF stages and could be the 400 Hz frequency commonly used. Signal tracing procedures then follow those previously given for other circuitry. Initially the signal is injected into the input to the last IF stage preceding the video detector. If a signal is observed on the scope, the generator is moved to the input of the IF stage feeding the last one. Again, if a signal is observed on the scope

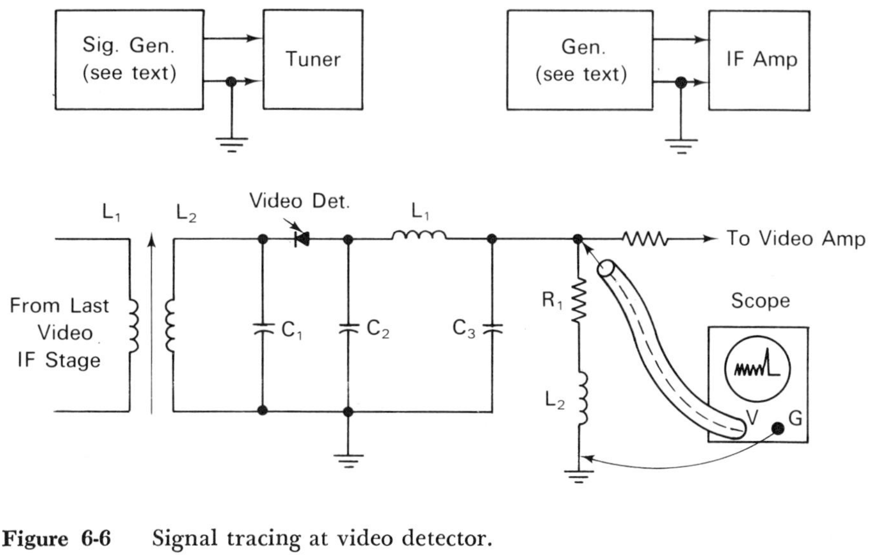

Figure 6-6 Signal tracing at video detector.

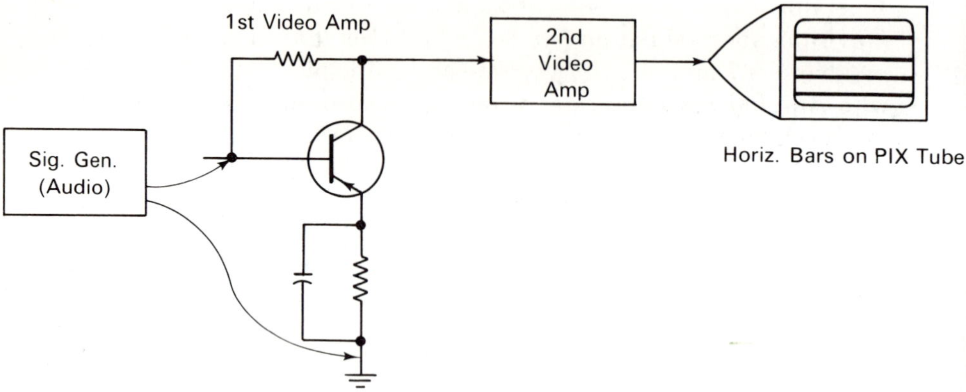

Figure 6-7 Video signal injection.

the generator is moved to another stage, thus working toward the tuner. Once no signal is observed, the inoperative stage has been localized.

If signal-tracing procedures are undertaken in the IF and tuner stages, it is assumed that the video amplifier stages that follow the detector are all functioning. Thus, in the absence of picture and sound, it could indicate the fault exists prior to the video detector, because a defect in the video amplifier stage may kill the picture but may not disturb the sound output. Similarly, lack of sound but the presence of a picture indicates faults in the sound IF stages only (following the video detector) and not the video amplifiers preceding the picture tube.

It is possible, of course, that lack of both sound and picture could mean that defects have occurred simultaneously in both the sound and picture amplifier stages following the detector. Thus, before signal tracing tuner and IF stages, the sound circuits (FM detector and amplifiers) should be traced. Also, a check can be made of the video amplifier stages by injecting an audio signal into the first video amplifier as shown in Fig. 6-7 to produce sound bars on the television screen. If no sound bars are observed the generator should be applied to the input of the second video amplifier.

As a quick check on the entire system, the signal generator can be dispensed with initially and a test made to see if any video signal is arriving at the video detector. With a strong station tuned in and a scope applied as shown in Fig. 6-6, the video signal should be visible. If the scope horizontal sweep is set at the 60 Hz rate the video signal's vertical sync pulses will lock in as shown in Fig. 6-8. The peaks and valleys between the sync pulses will vary constantly as the televised scene changes. Thus, the signal will not appear exactly as shown in Fig. 6-8,

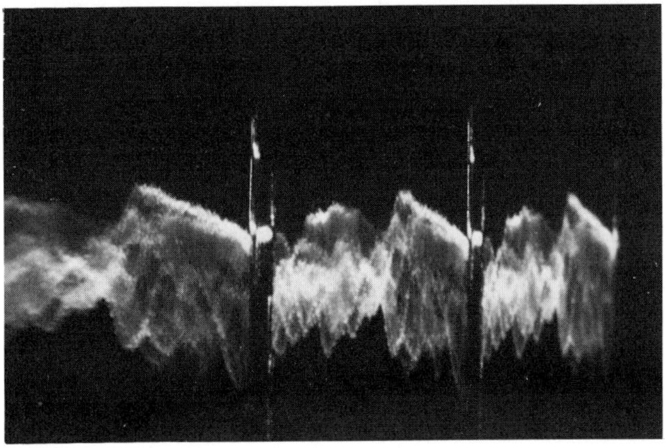

Figure 6-8 Video signal (60 sync rate).

but one similar to it will indicate that the tuner and IF stages are passing the signal to the detector. Hence, if no picture appears on the picture tube, the trouble is in the video amplifiers, contrast control, brilliancy control or the picture tube itself.

The oscilloscope may also be adjusted to lock in the video picture using the 1750 Hz rate of the horizontal sync pulses as shown in Fig. 6-9. Again, the signals displayed between pulses will vary as the scenes of the televised picture change. Since signal tracing is being undertaken, however, we are more interested in the presence of the picture rather than its nature. If amplitude appears below normal, then voltage and component checks must be undertaken, as well as alignment checks as detailed in Chapters 7, 10 and 11.

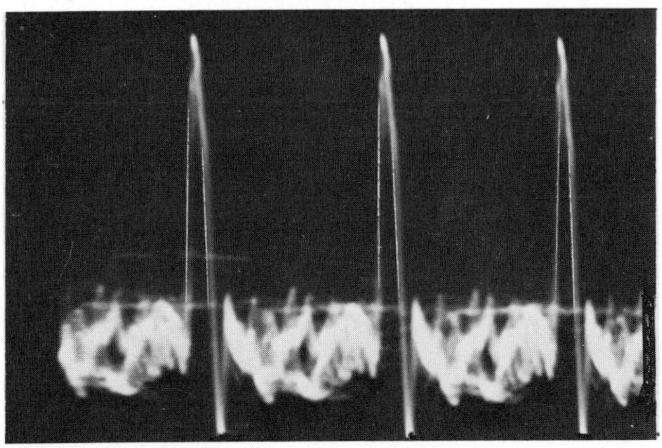

Figure 6-9 Video signal (1750 sync rate).

7

tracking and alignment for AM, FM and TV

7-1 TRACKING AND ALIGNMENT FACTORS

In the superheterodyne principle used in AM, FM and TV, the incoming carrier signal and sidebands are heterodyned (mixed) with a signal generated by an RF oscillator in the tuner. The result is the production of an intermediate-frequency (IF) signal containing all the sideband information of the original. The tuner resonant circuits in the mixer stage must *track* with the oscillator frequency to maintain the same difference frequency regardless of which station is being tuned. If an RF stage is present, it must also be tuned in coincidence with the mixer and oscillator. Adjustment to the tuning of these stages to maintain the proper difference frequency is known as *tracking* (see Figs. 5-8 and 6-3).

The IF stages of the receiver are tuned only to the single IF frequency produced by the tuner. The process of tuning each IF stage to this proper frequency is known as *alignment*. A properly aligned receiver achieves maximum performance and gain in the IF stages, plus proper bandpass for the group of signals to be handled. Thus, correct tracking and alignment procedures require a knowledge of the bandpass characteristics, frequencies and limits of AM receivers, FM receivers and TV sets, for each has frequency and bandpass requirements different from the others.

Tracking and alignment of a receiver are performed initially by the manufacturer and in most instances are fairly close to the precise frequencies to provide for the best reception for which the receiver is capable. On rare occasions it may be desirable to retouch the adjustments to obtain peak performance because tuning at the factory lacked precision, or because an inexperienced technician has misadjusted tracking and alignment. It is very unlikely that the receiver's circuits drifted off resonance appreciably, especially in the IF stages. Aging of transistors and tubes may alter resonant conditions slightly, particularly in the tuner, thus requiring adjustment. In all instances, however, tracking and alignment procedures should not be undertaken unless one is certain it is needed. Often tracking and alignment procedures are undertaken needlessly when the actual reasons for poor performance may be improper voltages, defective resistors or capacitors, or weak tubes and transistors in the tuner and IF stages.

The signal generators used for tracking and aligning wide-band stages (such as in FM and TV) require accurate settings and special characteristics. Their general features are included in this chapter. Manufacturer's service notes usually specify the step-by-step procedures recommended for tracking and alignment and should be consulted when such retuning is necessary. The data included in this chapter serve as a general guide for the basic principles involved and the special precautions necessary for achieving optimum results.

7-2 BANDPASS

The bandpass characteristics of AM and FM radio and TV differ considerably, hence each requires specific equipment for tracking and alignment procedures. The bandwidths for AM and FM are shown in Fig. 7-1. For the AM bandpass shown in Fig. 7-1A the 10 kHz total width means that sidebands extend 5 kHz above and below the carrier frequency. In AM two sidebands are created for each modulating signal. Thus, for a station operating at 1000 kHz and broadcasting a 1000 Hz tone, there would be an upper sideband at 1001 KHz, and another sideband below the carrier at 999 kHz, producing a bandwidth of 2 kHz. In areas where several stations are crowded together, interference can result by transmitting signals around the 5 kHz region (see Appendix C).

In FM the carrier is made to shift above and below its center frequency at a rate corresponding to the modulating signal. Thus, if an

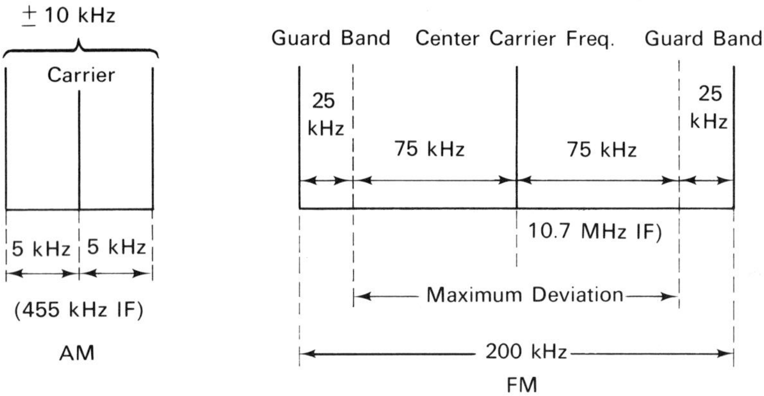

Figure 7-1 AM and FM bandwidths.

8 kHz note is used to modulate the carrier, it shifts 8 kHz above and below the center frequency. For a louder 8 kHz signal, the carrier shifts to a greater extent beyond the center frequency, but the rate of change still is 8 kHz. Many sidebands are produced, but only a few are significant in terms of sufficient amplitude. Still, with carrier shifting and multiple sideband production, a wide band is necessary for good reception. Thus, for the public-entertainment FM stations (88 to 108 MHz) the bandpass factors shown in Fig. 7-1B prevail. Note that 75 kHz is allocated each side of center frequency to leave room for sidebands and carrier swing. In addition, 25 kHz is provided at each side as guard bands to minimize the possibility of signals spilling out of their alloted bandpass and causing interference. Thus, the total bandpass for this type FM is 200 kHz.

Older receivers used intermediate frequencies that differed from the standard generally used in modern receivers. Thus, for AM receivers the 455 kHz frequency has become standard for the IF, while that for FM is 10.7 MHz. The FM sound bandpass accompanying television picture information is much narrower than the 88 to 108 MHz allocations, as described more fully later in this section.

The bandpass characteristics shown in Fig. 7-1 cannot be perfectly achieved in receivers because resonant circuits do not provide perfectly vertical sides. Thus, if a bandpass allocation is as shown in Fig. 7-2, from point A to B, a poor receiver alignment would result in the curve shown. Though the bottom portions extend to the bandpass limits, the actual bandpass is determined as the distance between two points on the graph, with each point designating a frequency at which the amplitude of the slope is 0.707 of the peak amplitude. These points are also termed *half-*

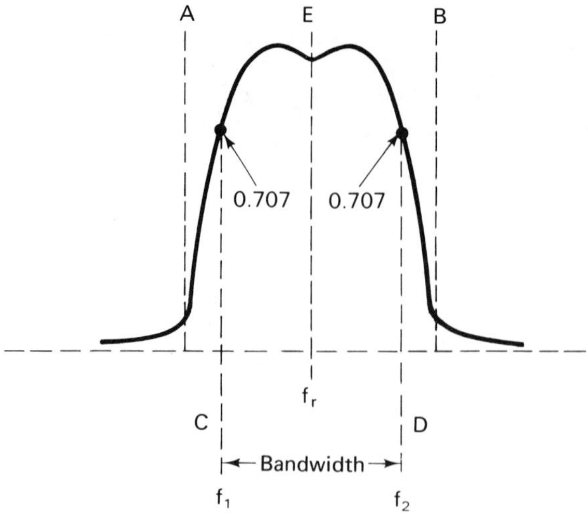

Figure 7-2 Half-power points determine bandpass.

power points. The f_1 and f_2 points, in relation to bandwidth, can be stated as:

$$\text{bandwidth} = f_2 - f_1 \tag{7-1}$$

Thus, if these points occur at 995 kHz and 1005 kHz, the bandwidth would be 1005 − 995 = 10 kHz.

The selectivity of a circuit is its ability to discriminate against unwanted signals. With a resonant circuit that has little sensitivity to signals above and below the bandpass, the selectivity is excellent, since the circuit selects only those signals desired. Selectivity has for its symbol Q and is related to the f designations in Fig. 7-2 as shown by the equation

$$Q = \frac{f_r}{f_2 - f_1} \tag{7-2}$$

hence

$$f_r = Q(f_2 - f_1) \tag{7-3}$$

and

$$\text{bandwidth} = \frac{f_r}{Q} \tag{7-4}$$

The dip in the curve at the resonant frequency (f_r) is a characteristic caused when two resonant circuits are tightly coupled. Sometimes another stage of amplification is tuned to the resonant frequency to smooth out the response. Also, several stages may be present where each is tuned to a slightly different frequency so as to obtain a wide bandpass with a fairly flat top as detailed in Sec. 7-6. The scope patterns will display the actual resonant curve present by using a signal generator capable of sweeping across the span of signal frequencies making up the bandpass as described in Sec. 7-3. Thus, a pattern such as shown in Fig. 7-3 may be obtained, where the wide band response decreases overall gain. The unequal response each side of center may be undesirable (as with AM and FM), or one side may be caused to slope deliberately (as with TV).

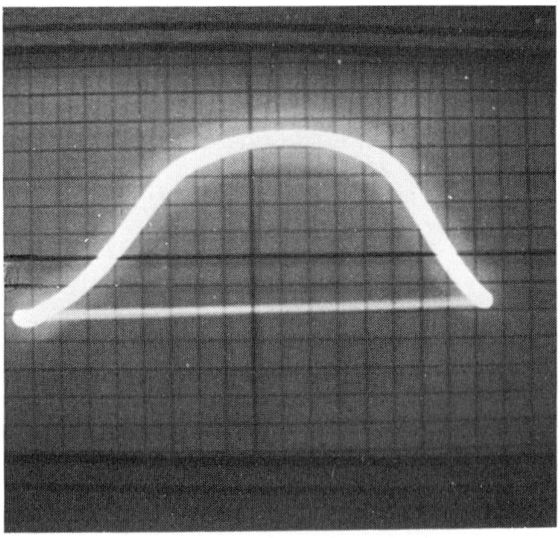

Figure 7-3 Unequal IF response each side of center.

In television broadcasting, two carriers are transmitted; one (which is amplitude modulated) carrying the picture information and the other (which is frequency modulated) furnishing the sound signals. The FM sound carrier is situated above the picture carrier as shown in Fig. 7-4 and the bandwidth is narrow compared to FM radio. With only 25 kHz deviation each side of center frequency, a maximum of 50 kHz bandpass is present. Thus for a given carrier power and modulating signal, reception will not equal that for FM radio.

The picture-signal information used to modulate the picture carrier ranges from below 60 Hz to about 4 MHz. If a 4 MHz signal were used to amplitude-modulate a carrier, two sidebands would be produced that would be separated by 8 MHz. Such a modulating signal would require an excessive bandpass. Consequently, most of the lower sidebands are suppressed at the station, and only a remnant (vestigial) portion remains at the low-frequency end of the picture signal span as shown in Fig. 7-5. This vestigial section can cause some interference to adjacent stations. Similarly, the sound carrier of one station can interfere with

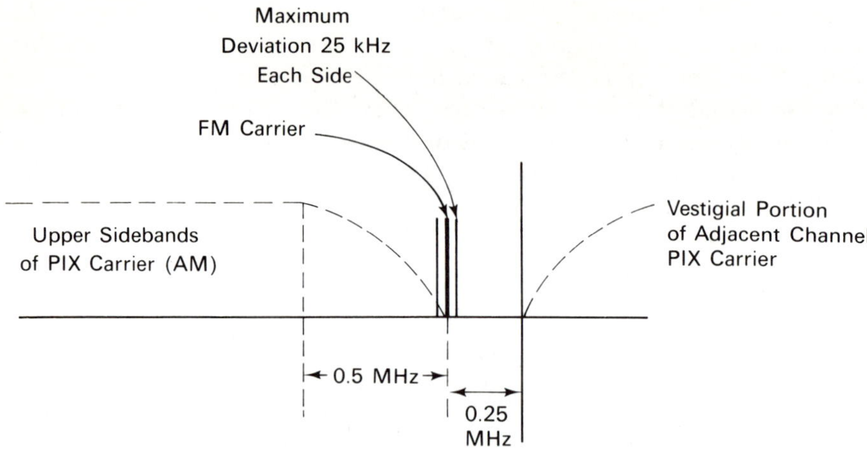

Figure 7-4 FM sound portion of television channel.

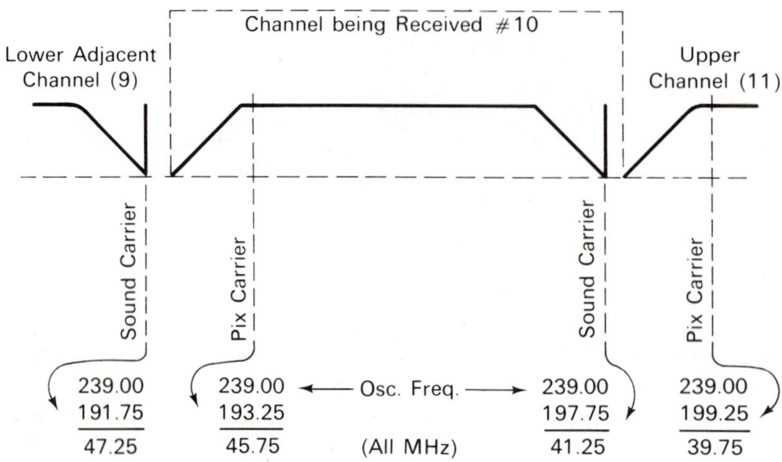

Figure 7-5 Carriers involved in TV tuner mixing.

an adjacent station. The closeness of the lower-channel sound carrier and the upper channel picture carrier to the station being received is shown in Fig. 7-5. (See Appendix C for carrier frequencies of other television stations.) To minimize interference, special traps are included in the television circuitry as shown in Fig. 7-6. These trap settings are usually checked during alignment procedures or when interference patterns are present (see Chapters 10 and 11).

These traps form series-resonant circuits to ground and provide a low-impedance path for signals to which they are tuned. The frequencies

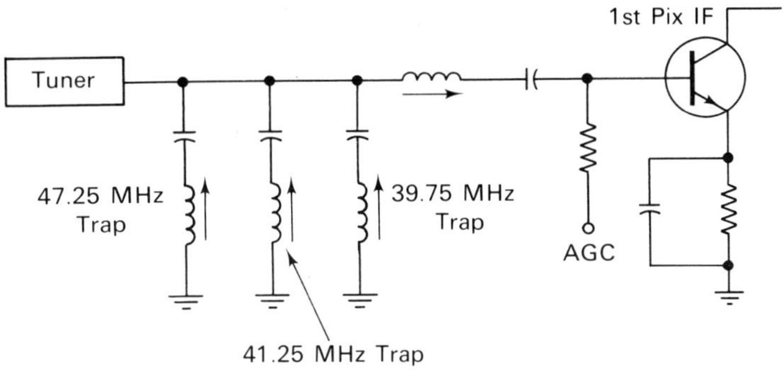

Figure 7-6 Typical TV traps.

for the interfering signals, as with the IF signals, remain fixed from the output of the tuner because the oscillator varies in frequency to maintain IF coincidence with the mixer stage. This can be seen in Fig. 7-5 where the IF signals are shown for Channel 10. For this station the oscillator frequency would be 239 MHz and the mixing process produces 47.25 MHz for the lower-channel sound, 45.75 MHz for the picture signal to which the station is tuned, 41.25 MHz for the sound signal and 39.75 MHz for the upper-channel picture signal. Had channel 9 been tuned in, the tuner oscillator would have been resonant to 233 MHz and this frequency signal, mixed with the various signals of channels 8, 9 and 10, would again produce the same IF signals as those in Fig. 7-5. (Note in Fig. 7-6 that one trap is for 41.25 MHz, which is the sound for the channel tuned in. It is necessary to diminish this signal to prevent the formation of sound bars on the picture-tube screen.)

During the mixing process in the tuner the relative position of the sound and picture signals reverses because the subtraction of the lower-frequency signals from the 239 MHz oscillator signal leaves a higher remainder (and a lower remainder for the higher-frequency signals). Consequently, taking into consideration this inversion process and the dips in response caused by the traps, we obtain a bandpass characteristic as shown in Fig. 7-7.

The picture IF signal is set down approximately 50 percent on the slope of the response curve as shown. This decreases the abnormal gain that would result around this frequency region because of the vestigial section of the lower sideband. Note the span is 4.5 MHz from the sound to the picture frequencies. In black-and-white receivers the bandpass may not be as wide, sacrificing some picture detail for higher gain and fewer amplification stages. For color sets, however, it is essential to have a wide bandpass to assure good color balance and fine detail.

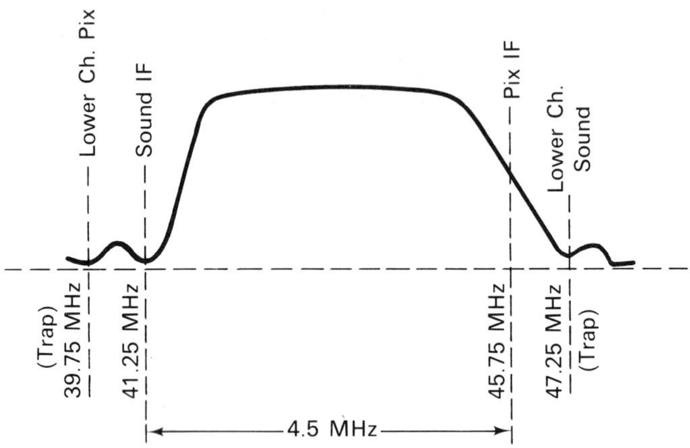

Figure 7-7 Frequency points on IF response curve of TV.

Again, as with FM, the display of the bandpass requires use of a sweep generator as subsequently discussed in this chapter. If the sweep is only over the bandpass, the results will appear as shown in Fig. 7-8. Additional sweep width would show the dips caused by the traps. While such a pattern displays the shape of the response curve, frequency points must be established for proper alignment of the IF stages, as discussed later in this chapter. Adjustment of the tuning cores of the interstage inductors will alter the bandpass accordingly and will change the slopes and flat-top expanse of Fig. 7-8. As pointed out in Chapter 4, however, the vertical and horizontal gain controls of the scope can change the

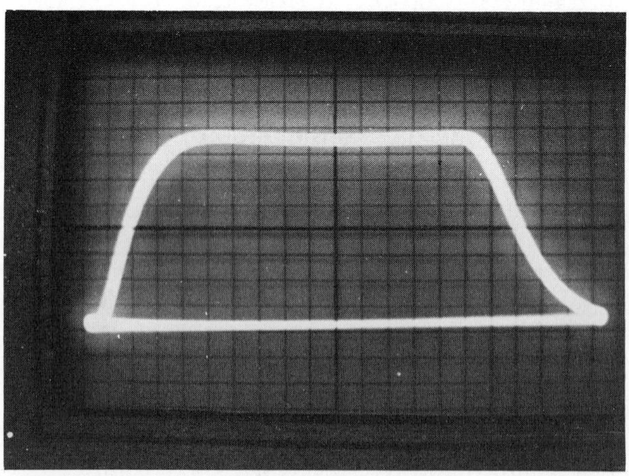

Figure 7-8 Typical scope representation of IF.

appearance of the waveform reproduced. Thus, for accuracy in alignment, frequency markers are essential (see Secs. 7-3, 7-6 and 7-7).

7-3 SWEEP AND MARKER GENERATORS

A sweep generator produces an output signal that constantly varies in frequency between two preset limits. The generator is modulated by another signal as in FM broadcasting, hence the amount by which the signal output changes its frequency as well as the amplitude of the signal produced can be adjusted as desired. The sweep generator is useful in producing an oscilloscope pattern showing the bandpass characteristics of a tuner or IF system. Thus, if the sweep generator output is applied to the first stage of an IF system and a scope probe placed across the detector load resistor the pattern shown in Fig. 7-8 appears.

The pattern is produced because the sweeping-frequency signal shows, at any instant, the relative gain of the IF system at that frequency. The sweep frequency starts below the resonant frequencies of the IF stages, so the gain is low and the height of the scope trace is down. As the sweep approaches the resonant frequencies, IF gain rises and the trace on the scope screen rises. Throughout the flat top of the wide-band IF system, the gain is constant and the scope pattern maintains its amplitude. Beyond the bandpass the gain is down and the scope trace declines.

The pattern produced by the sweep generator can be aligned properly only if the associated frequency points are known as were shown in Fig. 7-7. Such frequency spots can be identified by using another generator known as a *market generator*. This unit produces a single-frequency output signal and is essentially the type signal generator commonly used for signal-tracing and test purposes, as described in Sec. 6-1. When such generators are calibrated to produce accurate signals, they may be used as marker generators.

When the marker generator signal is injected with the sweep signal into an IF or tuner system, the marker produces a sinewave interference spot on the resonant curve as shown in Fig. 7-9. This spot, termed a *pip,* moves along the curve as the frequency of the marker is adjusted. The dial on the marker indicates the particular frequency at the point where the marker pip appears.

Modern sweep generators have built-in markers for convenience in tracking and alignment of wide-band FM and TV tuners and IF stages. Often push-button selection is provided for specific marker signals; these are controlled by individual crystals for accuracy. Although sweep-marker meter panels differ in placement of control knobs and jacks, representative controls are as shown in Fig. 7-10. The tuning is adjusted

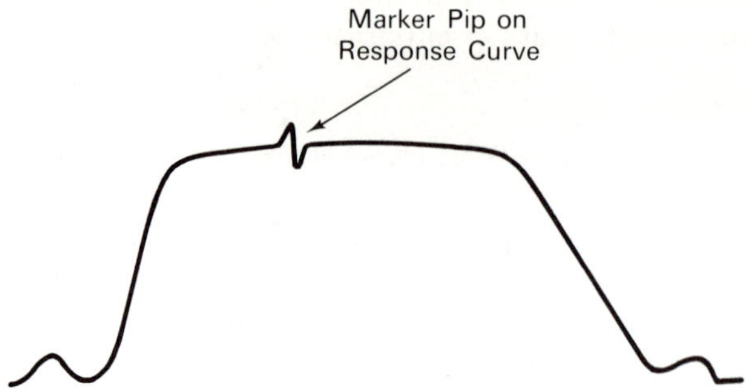

Figure 7-9 Marker pip as a frequency guidepost.

by the right-hand knob and the frequency around which the sweep occurs is displayed on the tuning dial. Output jacks are provided for the sweep signal and for the synchronizing signal for the oscilloscope. An input jack is also provided for injection of an external signal from a separate marker if desired.

The range-selector knob at the left selects the frequency span desired, from approximately 10 to 50 MHz for FM IF (standard 10.7 MHz) and 50 to 125 which takes in the FM band, 88 to 108 MHz, as well as channels 2 through 6 of television. Another range from 125 to 225 covers channels 7 through 13, and still another range extends to 900 MHz and is for VHF television (see Appendix C).

The output signal voltage is adjusted by the output control, which sets the μV level of the signal. Sweep width is regulated by the next control as shown, and may be variable over 15 MHz for wide-band sweeping beyond the bandpass curve. Another knob sets the amplitude of the marker pip as seen on the curve, and another control adjusts the width of the marker pip. Phasing between the sweep generator and the scope sweep is adjusted by the knob shown. Some sweep-marker generators have an input jack for plugging in the scope. In these the signal at the detector is picked up by a probe from the sweep generator and fed, with proper synchronization, to the scope.

With pushbuttons for the marker-frequency selection as shown in Fig. 7-10, predetermined frequencies are available instead of a continuously variable marker frequency as with variable-frequency signal generators. Since these marker signals are crystal controlled they are more reliable than those obtained from a variable-frequency marker, since the latter needs precise calibration for maximum usefulness.

Marker frequencies include those shown in Fig. 7-7, plus several along the flat top portion of the curve for aid in alignment. These additional markers may range from 42 to 45 MHz. Some manufacturers in-

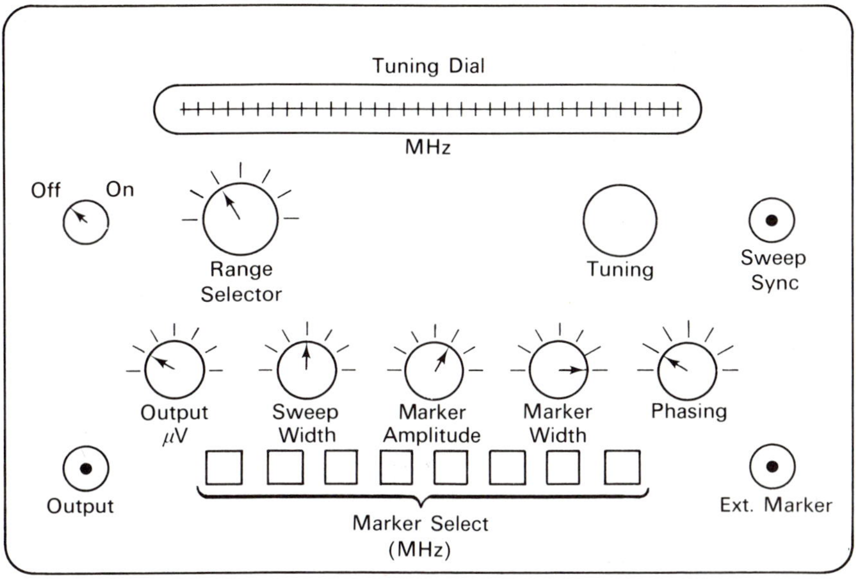

Figure 7-10 Sweep and marker generator controls.

clude crystal-controlled markers for the chroma carrier signals of color television for testing chroma bandwidth. These special marker signals include the 3.58 MHz frequency, plus a marker 0.5 MHz above and one 0.5 MHz below. Other special markers that may be present include RF video-carrier markers for several TV channels, and IF markers for FM, including such frequencies as 10.7 MHz, with another at 10.6 MHz and one at 10.8 MHz for bandpass alignment.

Some sweep-marker generators are available for exclusive usage on AM and FM, with no provisions for TV signals. While single-signal generators can be used for AM alignment, since the IF stages are not wide band as with FM, the sweep method is also useful because of the full bandpass characteristics displayed by the sweep and oscilloscope procedure.

As with most testing equipment, the features and versatility of sweep and market generators are directly related to their cost. Consequently the technician must relate his equipment needs to the costs involved, and endeavor to obtain units that will serve him best.

7-4 AM RECEIVER ALIGNMENT

When an AM receiver requires alignment and tuner tracking the symptoms include weak signal reception, higher than normal noise levels, and interference between stations. As IF stages become off resonance for the 455 kHz frequency, gain drops and hence the signal-to-noise ratio decreases. Off resonance of the IF stages also upsets the bandpass, increasing the possibility of undesired signals being accepted. If the tuner does not track properly, it may deliver the wrong-frequency signal to the IF stages, again degrading performance.

If the radio had been operating normally and suddenly developed one or more of the symptoms mentioned above, indications are that some component part (transistor, diode, capacitor or resistor) has become defective or totally inoperative. Thus, a sudden display of poor reception calls for other checks, and alignment should not be attempted until other possibilities have been ruled out.

It is possible, of course, that the alignment and tracking could be touched up slightly during servicing procedures to obtain peak performance. Once aligned, however, circuits may drift off resonance over a period of time as components age, but abrupt resonant changes are unlikely and could only be caused by a trimmer capacitor changing value or becoming disconnected from the primary capacitance of the resonant circuit.

Because of the narrow bandpass of AM radio compared with FM and TV, alignment and tracking procedures are comparatively simpler, requiring only a signal generator and a voltmeter. A sweep generator and scope could be used, of course, as with the alignment procedures for FM and TV. These instruments would permit direct visual observation of the bandpass curve. Generally, however, the single-signal generator and the voltmeter suffice for the standard AM radio receiver.

If tracking is attempted before IF alignment, it may require picking up the injected signal at the output of the tuner, particularly if the IF stages are badly out of alignment or defective. Because signal output is weak at the tuner a highly sensitive instrument is needed to get a reading. Thus, it is preferable to align the IF stages initially to permit a reading at the detector load resistor for both tracking and alignment procedures.

Initially the 455 kHz IF signal is injected into the resonant circuits coupling the last IF amplifier to the detector system, as shown in Fig. 7-11, at point 1. The voltmeter is placed directly across the detector load resistor as shown and left in this position for successive injections of the signal at other points shown. For each signal injection the appropriate inductor cores are tuned to obtain maximum reading at the detector load resistor. Manufacturer's notes, if available, should be checked for the recommended method of disabling the automatic volume control (AVC) system. With the AVC functioning, it would automatically reduce the gain as the circuits are tuned, and misleading readings may be the

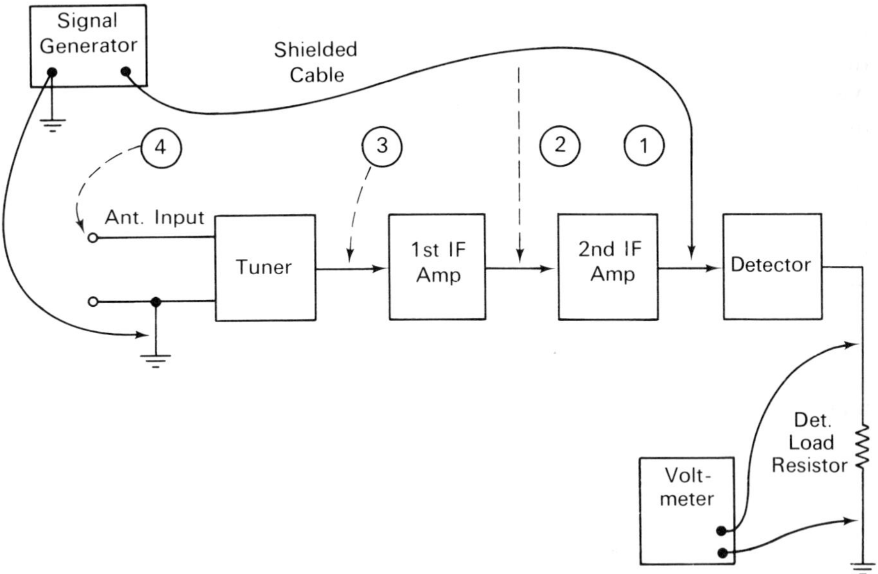

Figure 7-11 AM receiver alignment.

result. General practice is to apply a fixed bias to the AVC line and also disconnect the AVC feed from the detector circuit. Thus, the AVC bias remains fixed during the alignment.

Signal input from the generator should always be below the value that would cause transistor and circuit saturation. Too high a signal input would prevent observation of the results of tuning, since a saturated stage has a constant output (see also Sec. 7-7). As the stage is tuned and output levels increase, the output from the generator should be decreased accordingly.

Once the IF stages have been aligned, the signal generator is applied to the input of the tuner, as shown in Fig. 7-11, and checks made at the high-, middle- and low-frequency regions between 455 kHz and 1600 kHz. Trimmer capacitors are varied or other tuning adjustments made so that the *difference* frequency between the mixer and the oscillator is maintained at 455 kHz from the low-frequency end of the dial to the high-frequency end. This is checked by keeping the output of the signal generator at a constant level while changing its frequency. For each new frequency to which the generator is set, the radio is tuned accordingly and the output at the detector observed. Only minor voltage variations should be noted from one end of the band to the other.

Of importance in tuner tracking of AM receivers is the exact setting of the tuning dial pointer versus the proper oscillator frequency to produce the proper IF. This is checked by what is termed *rocking* the frequency tuning for several settings of the oscillator frequency. Assume, for instance, that we set the generator at 900 kHz and then set the tuning dial at this same frequency. While observing the metered output at the detector, the oscillator trimmer is adjusted very slightly to shift the tuning a little. Now the output at the detector drops. The radio tuning dial is now rocked (moved slightly to right and left) noting the output level. Often a higher output level will be obtained than originally. If not, tune the oscillator frequency slightly in the opposite direction and again rock the radio tuning. If greater output is obtained when the radio is tuned to 898 or 903 kHz, for instance, the dial pointer must be reset on the dial cord assembly to point to the 900 kHz dial markings to coincide with the 900 kHz input from the generator.

The foregoing rocking procedure can also be performed without instruments by using the signal from a station and listening for results from the speaker. If, for instance, a radio station is tuned in that has a frequency of 700 kHz, the tuning is varied around this frequency for several slight changes of the oscillator frequency. When a dial setting has been found that gives a maximum volume for a given oscillator frequency, the dial pointer is adjusted to read 700 kHz. Since a slight turning of an oscillator trimming capacitor can shift the oscillator frequency to a considerable degree, care must be taken to make only slight changes in the oscillator frequency for the rocking procedures.

Transistor replacement in the tuner or IF stages may alter resonance somewhat if the characteristics of the new transistor differ slightly from the old one. Perhaps, also, tracking and alignment had been changed to conform to the change in transistor characteristics as the unit aged. Even though characteristic changes in transistors are not as great over a period of time as was the case with tubes, a sufficient change might occur in which slight touch up of tracking and alignment may peak the performance again. Consequently, when new transistors are installed in the future, retuning may be necessary.

7-5 FM RADIO ALIGNMENT

Because of the wide-band characteristics of FM radio, proper tracking and alignment require use of a sweep generator with marker, and an oscilloscope. With these instruments the bandpass curve can be observed and as the various stages are tuned, the changes that occur in the overall bandpass curve can be observed. As discussed in Sec. 7-3, the marker generator produces a pip on the bandpass that serves as a road sign for ascertaining the frequency at that particular point.

As shown in Fig. 7-12, the sweep generator must be linked to the scope for proper synchronization of sweep between the two. With the output of the sweep applied to the base of the mixer transistor, as shown, two types of patterns may be observed in FM alignment. One is the standard bandpass curve and this is picked up at the last IF stage by a detector probe. If the particular system used includes a limiter stage preceding the detector, the signal can also be picked up there.

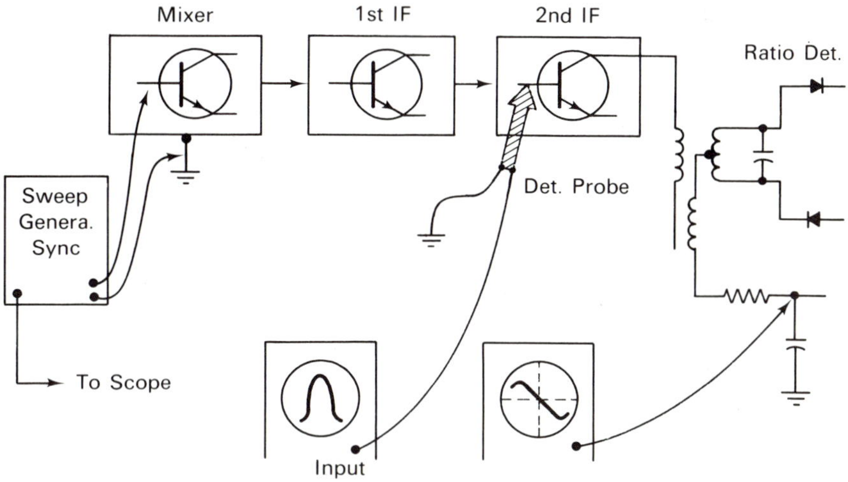

Figure 7-12 FM alignment checks.

When the signal is observed at the ratio detector (or discriminator detector) a curve is obtained showing relative gain above and below the center frequency of the carrier. Whether this curve slants to left or right depends on the phasing relationships of the individual sweep circuits or the generator and scope. Generally the curve is depicted as shown in Fig. 7-13 and resembles the capital letter S, hence is often referred to as an S-curve.

The S-curve displays the response at the detector to the shifting of the input frequency by the sweep generator. As the generator produces a signal with a decreasing frequency from center frequency, the detector voltage rises to a maximum amplitude as shown. With a shift of frequency in the other direction, the voltage drops to zero and rises again, but is of opposite polarity to the initial voltage.

Using the marker pip to establish the 10.7 MHz center-frequency of the IF, a pip appears as shown in Fig. 7-14A. The scope pattern is then shifted vertically and horizontally so the pip appears at the intersection of the *y* and *x* axis as shown. Now the symmetrical response at

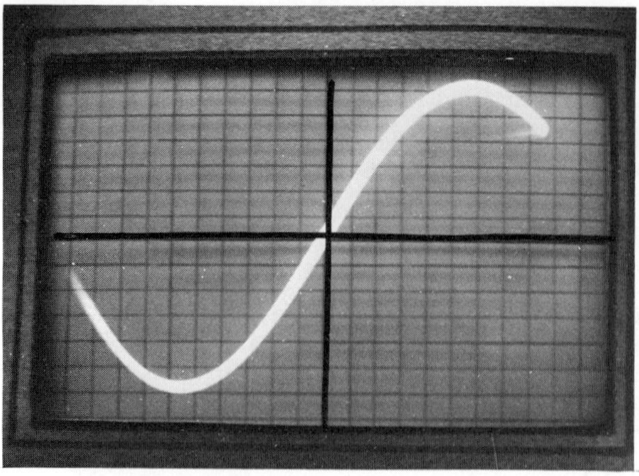

Figure 7-13 S curve (FM alignment).

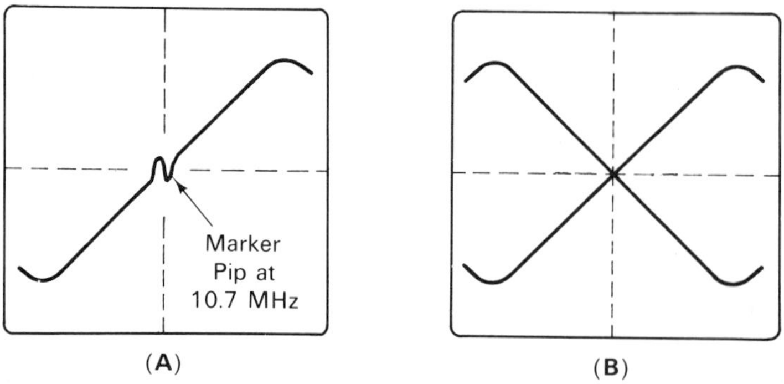

Figure 7-14 Marker point and double S.

each side of center frequency can be observed and the resonant circuits of the detector adjusted to compensate for any differences that may appear, such as more gain in the positive direction than the negative, or peaks reaching maximum amplitudes at different distances from the center 10.7 MHz point.

With some marker-sweep generators, a double S-curve can be obtained seen in Fig. 7-14B, where the center-frequency crossover point is automatically established. Since this crossover may not be at the desired 10.7 MHz point, however, the marker pip is used to find the exact frequency along the slopes of the S-curve and tuning adjusted to coincide with the true crossover point displayed. The single S-curve as in Fig. 7-14A is traced out when the sweep generator and the scope's horizontal sweep are synchronized sinewaves. Thus, as the sweep generator alters the frequency from below center to the high-frequency swing above center, the scope's sweep traces out the S-curve. On the return frequency sweep of the generator, the scope retraces the original S-curve from a positive high to a negative maximum value. If the scope uses a sawtooth sweep while the generator utilizes sinewave sweep, the maximum positive amplitude of the S-curve is abruptly swung to the left of the scope screen because of the rapid sawtooth retrace. The resultant is a double S-curve as shown in Fig. 7-14B.

Inductors used in tuned circuits in the FM region (88 to 108 MHz) have only a few turns and use larger diameter wire than coils at AM. Hence there is little that can change their inductance once they are tuned. As with AM discussed in Sec. 7-4, it is unlikely that alignment in an FM receiver would change suddenly to any marked degree. Thus, when symptoms of poor gain and interference are present the circuits should be tested for other troubles. Only when one is sure that the faults are caused by misalignment should realignment and tracking be undertaken. The manufacturer's instruction regarding alignment should be consulted if available. Also, commercial sweep generators and other alignment equipment have complete instructions for proper set up of tracking and alignment and should also be consulted (see also Sec. 7-7).

Specifically, a manufacturer may recommend that a stereo receiver be switched to mono during alignment, with the AFC switch (if present) turned off. Similarly, specific instructions are usually given regarding how any automatic bias is made inoperative and a form of fixed bias applied. Voltage levels to be expected and step-by-step instructions may also be detailed.

Often, in modern receivers, the IF transformers are replaced with fixed-frequency crystals or ceramic filters for assuring precise bandpass characteristics. These devices contribute to rejection of undesired signals because of their accurate characteristics. These filters are not necessarily resonant at exactly 10.7 MHz, but may vary between 10.6 and 10.8 MHz for various receivers. Thus, if a particular receiver uses crystal filters that are matched to produce 10.68 MHz, the mixer-oscillator circuitry of the tuner must then provide this frequency for proper alignment.

The partial schematic shown in Fig. 7-15 illustrates the circuitry associated with the crystal filter. Here, the oscillator-mixer transistor has its output applied to a resonant circuit composed of L_1 and L_2. The adjustable-core tuning slugs in the primary and secondary of this interstage transformer must be adjusted so resonance is obtained with the *exact* frequency of the crystal filter. Thus, if a sweep generator has a frequency deviation above and below 10.7 MHz, the operation of the crystal filters cannot be checked accurately unless they are close to this frequency.

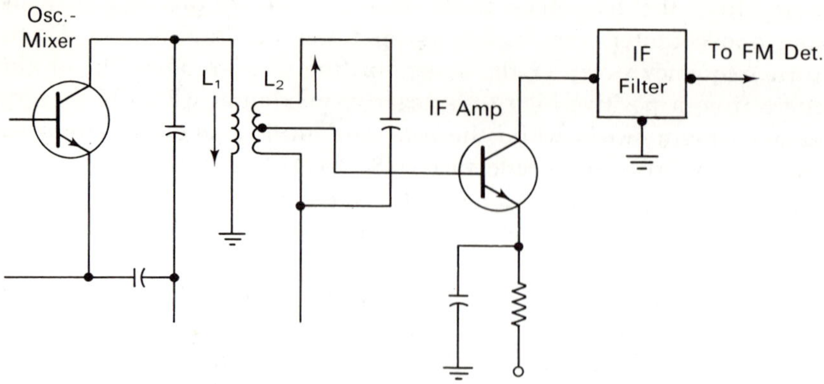

Figure 7-15 Crystal-filtered IF stage.

An accurate variable-frequency marker generator or other single-signal type can be used to determine the exact frequency of the crystal filters. Some sweep-marker generators specifically designed for FM service have a provision for rocking the center 10.7 MHz frequency to determine at what frequency maximum output is obtained. Once this is found the transformer primary and secondary (L_1 and L_2) of Fig. 7-15 are tuned to this frequency to provide for maximum IF amplifier gain.

For tracking, the signal generator is coupled to the tuner input, using isolating capacitors if required to prevent dc coupling. Sometimes a resistance network is recommended for impedance matching purposes. Tracking is checked at the low 88 MHz end, at the middle range around 98 MHz, and also near the 108 MHz end. The generator is adjusted for the sweep range desired (usually 75 kHz sweep each side) and the oscillator trimmer capacitor adjusted for maximum output. Again, the output from the signal generator must always be kept below the level at which limiting and saturation might occur (see also Sec. 7-4).

Stereo FM receivers differ from the mono types by the circuitry that follows the FM detector, as described more fully in Chapter 9. Thus, tuner tracking and IF alignment factors are the same for either. With the stereo receiver, however, it may also be necessary to undertake alignment of the stereo decoder circuitry; this subject is also covered in Chapter 9.

7-6 TV ALIGNMENT FACTORS

As shown in Fig. 7-7, the wide-band resonance curve for the IF stages of a television receiver is not symmetrical each side of center as is the case with AM and FM. In addition, the presence of the necessary traps (Fig. 7-6) places various dips in the curve, making for a much more complex set of characteristics than with radio receivers. As with FM alignment, however, only a sweep generator and a scope are required to observe the full bandpass characteristics. Hence, the basic set up is as shown in Fig. 7-16.

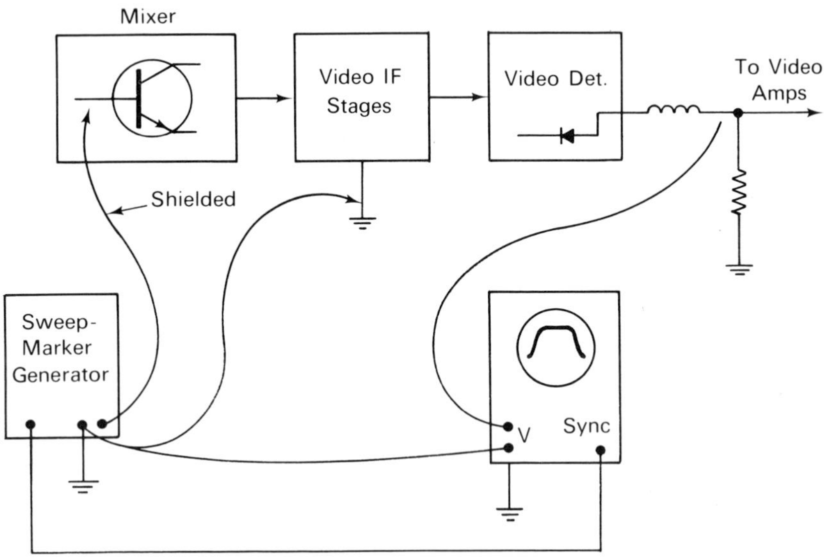

Figure 7-16 Video IF alignment.

It is worth emphasizing again that indiscriminate alignment should never be undertaken unless the technician is fully familiar with alignment procedures and understands how to test for the status of the bandpass before attempting adjustments. As with AM and FM, it is highly unlikely that alignment alters suddenly to a marked degree where it would cause picture interference, weak contract, poor colors or poor detail. Some of these problems may be caused by defective transistors and circuit components, and others may be the result of misadjustment of the focus controls, high-voltage section and so forth (see Chapters 10 and 11).

Marker pips are essential in alignment of television receivers, since we must know that traps are adjusted at the proper frequencies, and that the picture carrier is down on the 50 per cent point on the slope. As mentioned with AM and FM alignment procedures, it is useful to have the manufacturer's recommended procedures for tracking and alignment, since their detailed information furnishes the specifics related to that particular chassis and are often not applicable to other receivers.

The manufacturer's instruction manuals, schematics, and service notes assure proper step-by-step procedures and also discuss coupling factors, bias adjustments and control settings for the different procedures. Without such data (plus the instruction manuals accompanying the equipment to be used) the technician should not undertake alignment by following generalized rules. The latter, as given in this chapter, serve as a guide in avoiding some of the pitfalls that may be encountered in the complex procedures sometimes called for in tracking and alignment.

If the set-up shown in Fig. 7-16 indicates improper trap adjustments the trap tuning arrangement shown in Fig. 7-17 must be used. Here a single-signal generator (accurately calibrated) or a marker generator with crystal guide points is applied to the tuner input. Each trap requires the proper frequency signal from the generator. The appropriate trap is then tuned for a *minimum* reading from the voltmeter (since we are trying to trap out the entry of these signals into the IF system). After the traps have been adjusted the sweep generator and scope set-up of Fig. 7-16 can again be used to check the total response and to make sure the proper trap dips are present on the curve as required. Some marker-sweep generators provide for a tilting of the marker pip for greater visibility when shown on sharp slopes. This is helpful to localize the proper frequency of the trap, particularly when the trap is not tuned properly and the trap frequency pip would appear on the steep sides of the curve dip.

In some receivers a wide-band IF circuit is obtained by closely coupling primary and secondary windings of the interstage transformers (L_1 and L_2 of Fig. 7-15 for instance). A wide bandpass can also be obtained by tuning successive IF amplifier stages to different frequencies

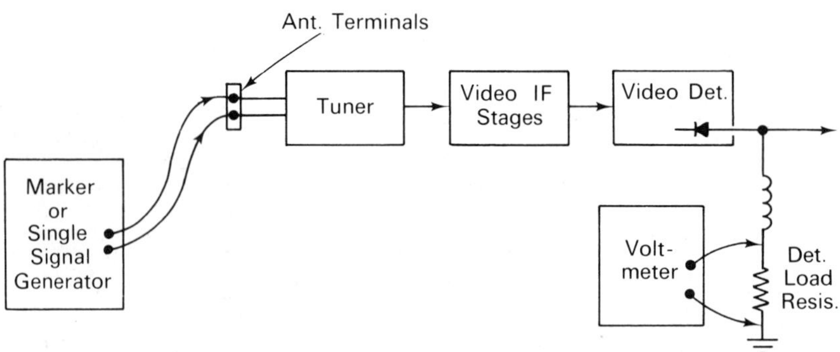

Figure 7-17 Trap tuning and alignment.

around the bandwidth required so that the sum of the individual narrow band-pass curves provides the total bandwidth needed. This system is termed *stagger* tuning. Such a system can be aligned with the single-signal generator and voltmeter method shown in Fig. 7-17. Since single frequencies are involved, as with the traps, the signal generator is set to the particular frequency of a particular IF stage and the resonant circuit of that stage tuned for maximum output. The generator is then set to the frequency of the next stage and that stage is tuned for maximum output. This procedure is followed for each individual tuned circuit involved.

Once the stagger-tuned stages have been aligned, the total band-pass characteristic can, of course, be viewed as a final check by using a sweep generator and a scope, connected again as shown in Fig. 7-16. As with all alignment procedures, it must always be remembered to keep the output of all signal generators below the point at which circuit limiting of the signal occurs. At saturation the output signal level remains constant and tuning procedures do not show output changes. (See also Sec. 7-7 for additional precautions relating to general alignment and tracking procedures.)

In television receivers both the picture IF of 45.75 MHz and the sound IF of 41.25 MHz are mixed again at the video detector circuit to produce a second sound IF of 4.5 MHz. Because the picture and sound IF have fixed frequencies, the heterodyned resultant (4.5 MHz) always remains at that frequency. This signal is amplified by one or two stages of IF amplification before detection (see Chapter 10). Alignment simply consists of tuning the one or two resonant circuits involved for maximum sound output. A signal generator can be used to produce a 4.5 MHz signal for testing and alignment purposes.

7-7 GENERAL ALIGNMENT PRECAUTIONS

As stressed in Sections 7-4, 7-5 and 7-6, alignment should not be undertaken unless one is sure that poor reception is not caused by defective components. Once tracking and alignment are indicated, the service notes for the receiver should be consulted if available. Also, the instruction manuals accompanying the generators and scopes should be consulted for guidance in proper hook-up procedures for tracking and alignment. Tracking and alignment involve high-frequency signals, particularly for FM and TV, and hence even small capacitance effects between leads can upset circuit function sufficiently to cause problems. Therefore, test leads should be as short as possible, and shielded cables are necessary for scopes and generators.

Generators, scopes, markers and the circuits being processed should all have a common-ground bond. A floating ground (a common circuit not linked to others) can produce hum interference and lack of a proper signal return circuit, and also can introduce some shock hazards on occasion. Good grounding plus adequate shielding will also minimize the pick up of undesired signals from circuit oscillators, which often radiate some signal energy even though shielding may be present.

Check for recommended usage of matching pads or isolating capacitors as shown in Fig. 7-18. Sometimes a 75 Ω generator cable should be matched to a 300 Ω receiver input by use of a matching resistor network, as shown in Fig. 7-18A, to obtain better signal transfer. Sometimes the recommendation is that an isolating capacitor be placed between the generator probe end and the input to the IF system for isolation of dc potentials such as bias voltage, which could be shunted by the probe (see Fig. 7-18B). If the scope picks up undesired signals it may be necessary to use a small shunting capacitance at the input (Fig. 7-18C) to provide a low-reactance path for undesired high-frequency signals, but still pass the low-frequency signals we want displayed. The capacitor value will depend on the type of interference present and can range from 0.02 μF to 0.001 μF.

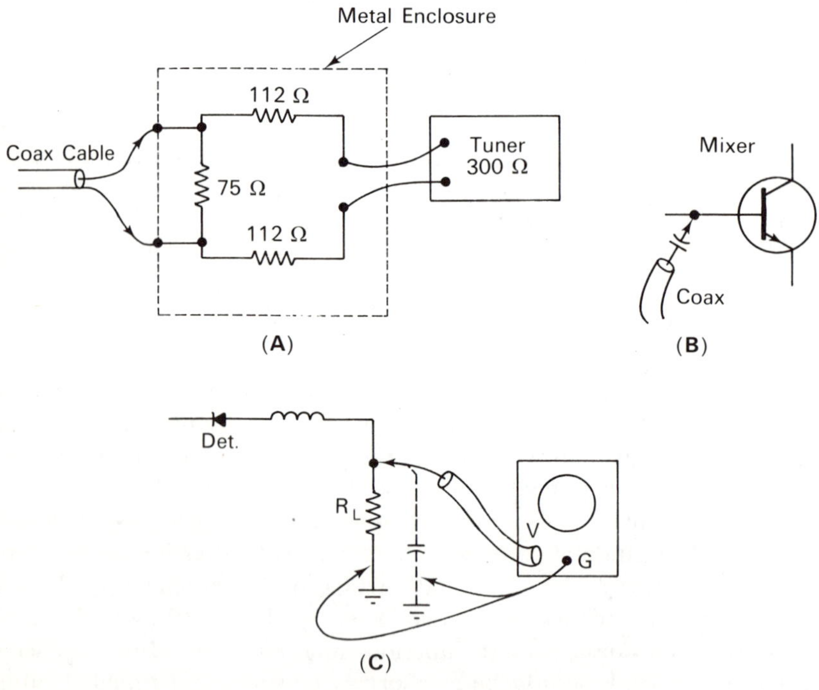

Figure 7-18 Alignment aids.

Keep the output from any generator below the level at which circuit limiting or saturation occurs. Circuit overload by a generator signal results in the maximum signal transfer and makes alignment or tracking impossible. Adjust the gain control of the generator and note the results on the output voltmeter or scope. When the output from the IF system remains constant for an increase in the gain control setting of the generator, overload has occurred. The generator output should be held below this point. Reduce generator output for each tuning change that is made, because as the circuits are brought into alignment more signal energy will pass through.

While the common grounding of all equipment is essential, precautions must be observed when a "hot" chassis is involved. When the power mains are rectified without an intervening transformer, the chassis is connected to one side of the ac line and can present serious problems by causing short circuits. If in doubt, check for possible ac voltage between the chassis in question and other ground systems.

Most equipment, even though fully transistorized, should be left on for five to ten minutes before being put to actual use. This will minimize the possibility of signal-frequency drift during the warm-up period.

In TV alignment check the proper tuning of the traps, even though the receiver is used in a signal area where adjacent channel reception cannot be obtained. Improper setting of traps may result in diminishing signals that are part of the required response curve.

Be certain to check the AGC or AVC bias adjustments needed for alignment and tracking purposes. Usually the manufacturer's service notes make recommendations for such adjustments. AGC and AVC automatically increase RF-IF stage gain and hence interfere in obtaining proper output readings during tracking and alignment.

In all alignment procedures, whether AM and FM radio or television, as the gain of the stages is increased by the tuning procedures, the scope pattern picked up at the output (or the voltmeter reading at the detector) increases. If the receiver was significantly out of alignment, the scope pattern may overreach the tube face as the set is tuned nearer the proper alignment frequency. The immediate tendency is to reduce the vertical scope gain to bring the pattern down where it is completely visible again on the scope face. Similarly, if a voltmeter is used, the tendency is to switch the meter to a higher scale reading to stop the meter needle from going off scale. What should be done, however, *is to decrease generator output until the proper level is obtained* on either the scope or voltmeter, whichever is in use. As mentioned several times before in this chapter, overload of the stages must be avoided by keeping the generator output below the saturation point. At saturation the bandpass response on the scope appears perfectly flat and hence gives the false impression that the response is perfect. Actually, with decreased generator output, the true response may show evidence of requiring much more tuning adjustment.

For television receiver alignment and tracking, whether black-and-white or color, it is good practice to set all traps at the right frequency initially. Next, the video IF and sound IF stages can be aligned. Once these stages have been tuned satisfactorily, the RF and mixer stages are tracked to produce the proper intermediate frequency signal. Finally, the oscillator frequency can be checked and trimmers set for obtaining the best picture and sound at approximate center of the fine-tuning control setting.

In color receivers other resonant circuits are present (3.58 MHz oscillator, bandpass amplifier, etc.) and these adjustments are covered in greater detail in Chapters 10 and 11.

8

solid-state and tube testing factors

8-1 SYMBOLS AND TERMS

Transistors and other solid-state units are manufactured by processed elements that form semiconductors. Such semiconductor segments termed *zones* have either a positive or negative characteristic and when properly combined form triode (or tetrode) devices capable of amplification, signal generation and signal processing in a fashion similar to the vacuum tubes used so extensively for many years.

When *p* and *n* zones are combined they form diodes used for rectification and detection, since such junctions exhibit unidirectional characteristics. If a junction is formed of a *p, n* and another *p* zone as shown in Fig. 8-1A, the device is termed a pnp transistor as opposed to the *npn* type, which is also shown. The three terminal wires from the transistor are the emitter, base and collector wire, respectively. (Generally, the base compares to the grid of the vacuum tube, the emitter to the cathode and the collector to the anode.) Symbols for the two junction type transistors are in Fig. 8-1B. Note the arrow representation for the emitter points toward the junction line for the *pnp* but away from it for the *npn*. This is an important factor in transistor testing and voltage-

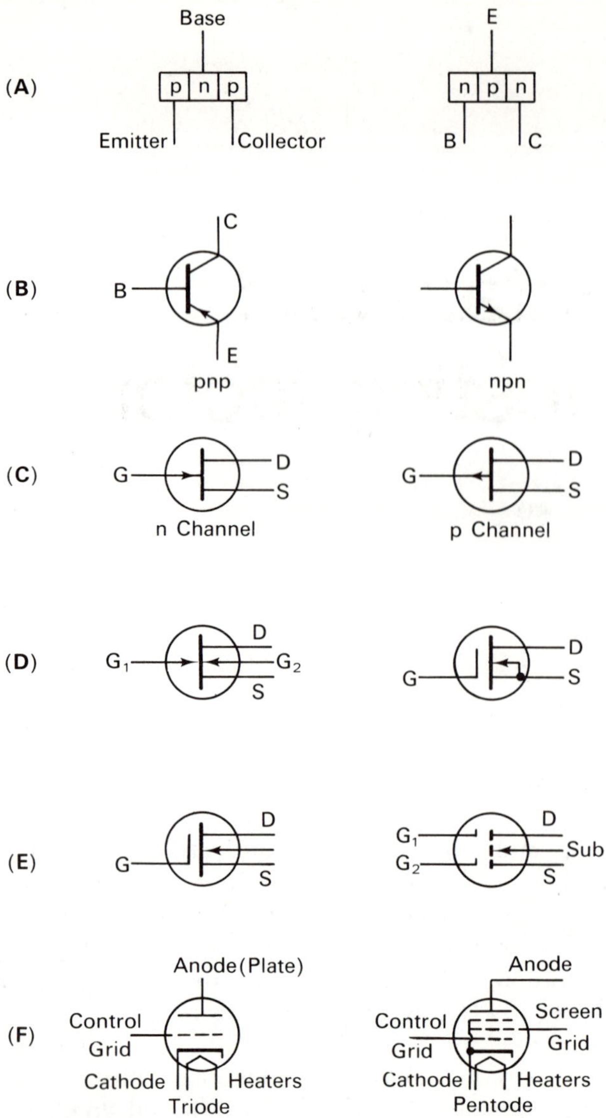

Figure 8-1 Transistor and tube symbols and terms.

reading checks, because battery polarities are different for the two types, as detailed in Sec. 8-2.

Another semiconductor device is the field-effect transistor (FET). The FET is sometimes called a *unipolar* (or *unijunction*) transistor because it has only a single junction. A better designation is *junction field-*

effect transistor (JFET). When the FET is manufactured from metal oxide it is called *metal oxide semiconductor field-effect transistor* (MOS-FET) and is termed an *insulated-gate field-effect transistor* (IGFET).

Compared to the transistor base terminal, the FET lead is termed a *gate*. The FET equivalent of the collector is the *drain* terminal, and the *source* terminal compares to the emitter of the junction transistor. As with the two transistor types, there are polarity differences with the FET as shown in Fig. 8-1C. When the gate arrows points toward the junction line, the FET is an *n*-channel type. As is also shown, when the arrow points away from the junction line it indicates a *p*-channel FET. Additional FET symbols are shown at D and E, including dual-gate types.

Two tube types are shown at F for reference; a triode and a pentode. The control grid has a basic function similar to the base and gate of semiconductors, because a small amplitude signal change between grid and cathode causes an amplified version to appear between anode and cathode. For the pentode the screen grid, by being made positive in polarity, accelerates the electron beam within the tube and increases gain and power-handling capabilities. The additional grid between the anode and screen grid is usually connected to the cathode within the tube, though on occasion it is brought out to one of the tube prongs for special applications. This additional grid is known as the suppressor grid since it stops secondary-emission electrons that may bounce off the plate and be attracted to the screen grid. By presenting a negative field between plate and screen, it repels the negatively charged electrons.

The heaters have a voltage applied to them to produce sufficiently high thermal conditions to heat the cathode so it will liberate electrons. These, being of negative polarity, are attracted to the positive anode. The disadvantage of tubes compared to transistors is the need for separate voltages to heat the filaments and the high voltages necessary for anode and screen grid (which often ran as high as 400 V in high-power audio amplifiers). Transistors, on the other hand, run cooler, consume less energy, are more compact and have a much longer life than the vacuum tube.

8-2 BIAS METHODS

The basic circuit for an amplifier using a *pnp* transistor is shown in Fig. 8-2A. For most conventional amplifiers of this type the emitter is placed at ground and hence the circuit is termed a *grounded emitter* or *common emitter* circuit. Because the emitter and base form a *p-n* junction, a positive voltage is applied to the emitter with respect to the base, hence the base is negative with respect to the emitter. When

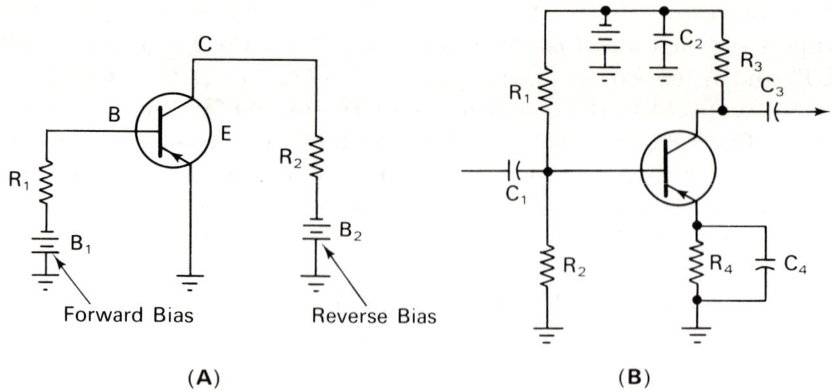

Figure 8-2 Common emitter circuitry (pnp).

battery polarities coincide with the junction (p-n) polarities it is called *forward bias.*

The voltage at the collector (with respect to emitter) is biased in reverse, that is, the *p*-zone of the collector is biased *negatively* with respect to the emitter. Resistor R_1 has applied across it the incoming signal to be amplified, and the output signal develops across R_2 from which it is applied to another circuit for additional processing, or to a speaker, or recording head.

The circuit shown in Fig. 8-2A is the manner in which the basic circuit is usually illustrated to show bias relationships. Actually, however, the practical equivalent has more components, as shown in Fig. 8-2B, and a single battery is used to provide for both the forward bias between emitter and base, and the reverse bias between collector and emitter. Capacitor C_1 couples the signal from the previous stage to the base input and across resistor R_2. Resistors R_1 and R_2 form a voltage divider to supply the base with the necessary forward bias. Since the negative terminal of the battery connects to the top of R_1, electrons flow through R_1 and R_2 to ground (and the positive terminal of the battery). The direction of electron flow places a negative polarity at the top of R_2, supplying the necessary bias.

Capacitor C_2 provides a bypass effect return path for the signals developed across R_3 to ground (and emitter) circuit. Capacitor C_3 couples signal energy to a subsequent circuit or unit. Resistor R_4 has a stabilizing effect on bias and the transistor's operating point, while C_4 returns the emitter circuit to ground because of the shunting effect of the capacitor's low reactance. Thus the emitter is kept at signal ground, even though R_4 would otherwise raise the emitter above ground.

Another grounded emitter is shown in Fig. 8-3A, though here an *npn* transistor is used instead of the *pnp* shown in Fig. 8-2. Note that forward bias is still applied between emitter and base, but since an *np* junction is involved, the base is positive with respect to the emitter. Similarly, the reverse bias between collector and emitter has opposite polarities than was the case for the *pnp* transistor.

An *n*-channel FET circuit comparable to the grounded emitter is shown in Fig. 8-3B. Note that the voltage is negative at the gate input with respect to the grounded source (also called common source). For the drain side a positive potential is applied, making the source negative with respect to the drain. If a *p*-channel FET were used in the same circuit of Fig. 8-3B, both battery polarities would be the reverse of those shown.

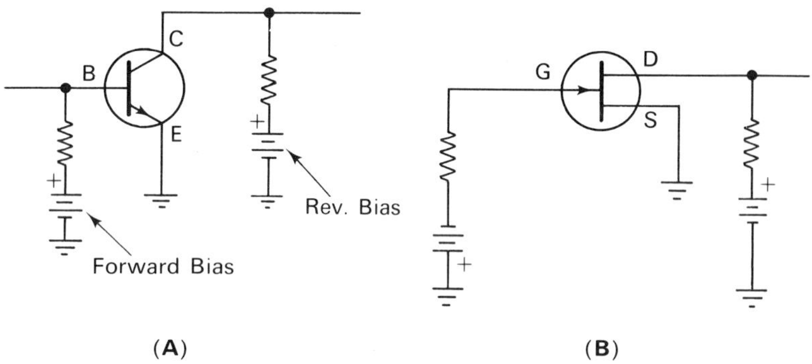

Figure 8-3 Grounded emitter and source circuits.

Other circuit types are shown in Fig. 8-4. The common gate FET system is illustrated in Fig. 8-4A. Here the gate circuit is at ground, with the input signal applied between source and gate and the output signal derived from R_2. In practical circuits of the type shown in Fig. 8-4 the batteries would be bypassed as shown in Fig. 8-2B. Also, the gate in Fig. 8-4A could have a series resistor with a shunting capacitor but the gate would still be at signal ground because of the capacitor's bypass effect.

The equivalent circuit for the common gate is the common-base circuit shown in Fig. 8-4B. Here an *npn* transistor is used, requiring battery polarities as indicated. Thus we still have forward bias between emitter and base and reverse bias between collector and base at the output. The common gate and common base circuits are sometimes used in RF amplifiers to provide better isolation between input and output circuits.

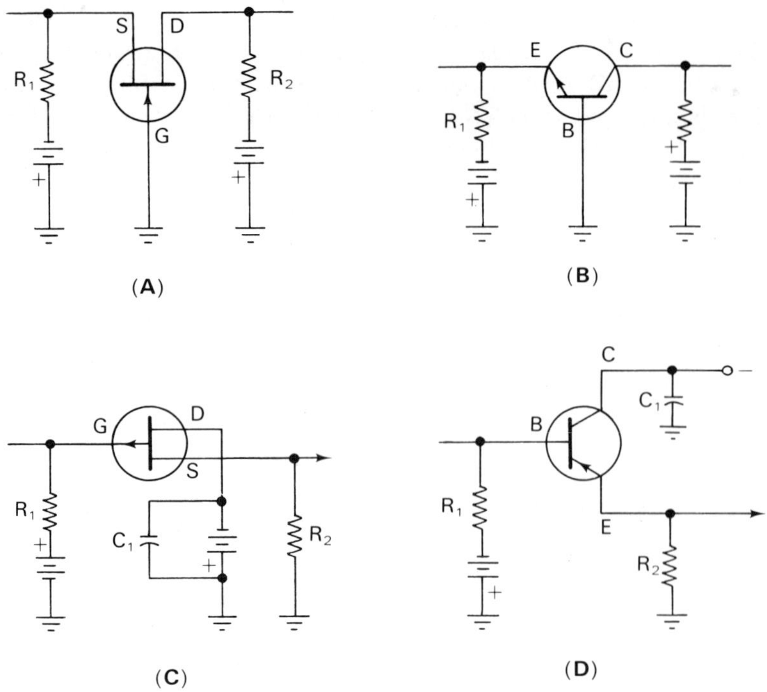

Figure 8-4 Common gate, base, drain and collector circuits.

The circuit shown in Fig. 8-4C is the common drain type, because C_1 has a low-reactance and shunts signals around the battery, thus placing the drain at signal ground. The output is obtained from the source, across R_2 to ground, hence from between the source and drain. The common drain compares to the common collector shown in Fig. 8-4D and to the tube-type common anode shown in Fig. 8-5C. These circuits are also termed *source follower, emitter follower* and *cathode follower,* so named because the output signal *follows* the phase of the input signal. (With the other circuits the output signal is opposite in phase to the input.)

For the circuit in Fig. 8-4D the collector is effectively placed at *signal* ground (while preventing grounding of the dc) by capacitor C_1. The input signal is applied between gate and ground across R_1 and the output develops across R_2 as shown. Such circuits are often used for proper impedance matching between circuits.

Basic tube-type amplifier circuits are shown in Fig. 8-5. A triode audio amplifier is shown in Fig. 8-5A, with bias provided by resistor R_2. As electrons flow through this resistor (current flow through the tube) a potential is set up as shown, with the cathode made positive with re-

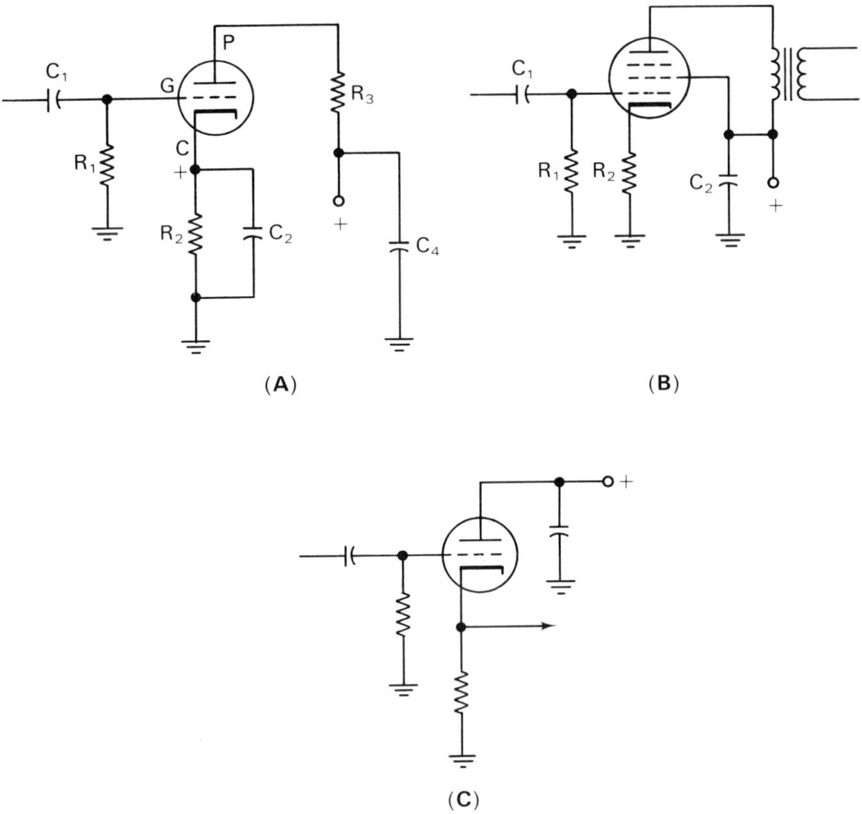

Figure 8-5 Basic tube circuits.

spect to ground. Since the grid is grounded by R_1, the grid is made negative with respect to cathode. The output signal is obtained from across R_3, and thus from between the anode and ground (because C_4 places the bottom of the resistor at ground, the same as the cathode).

For a pentode, the circuit shown in Fig. 8-5B is often used for an audio output circuit with a transformer feeding the voice coil of a speaker. Resistor R_2 can be bypassed, but sometimes the capacitor is omitted to improve fidelity at a sacrifice of some gain. (Since a capacitor has a lower reactance for higher signal frequencies it actually presents an uneven bypass effect and can hinder the obtaining of a flat response.) As shown, the screen grid is bypassed (C_2) to prevent signals from developing in it and thus reducing output gain. The circuit in Fig. 8-5C, already mentioned, is the cathode follower, which is characterized by a high input impedance and a low output impedance, thus forming an equivalent step-down transformer.

8-3 BASIC BIAS CHECKS

Basic measurements for bias and other voltages were shown earlier in Fig. 3-9. Similar voltage checks apply to the transistor circuits shown in this chapter, and to the tube circuits of Fig. 8-5. For tubes, a voltage check across the cathode resistor will show the amount of bias existing between the grid and cathode. An alternate method for checking tube bias is to place the negative voltage probe at the grid and the positive voltage probe directly at the cathode. (In a Class A amplifier no grid current flows, hence there would be no current through the grid resistor and no voltage drop across it. Thus, readings should not be taken from grid to ground for bias. A routine check of this sort can be made, however, to see if any positive voltage appears at the grid. If so, the tube may be defective or capacitor C_1 may be leaky and passing dc to the grid from the previous stage.)

In making bias checks on transistors and FET units, consideration must be given to the relative voltages appearing at the terminals. For instance, Fig. 8-6A shows an *npn* transistorized amplifier with typical manufacturer's notations of the voltages to be expected. Since this is an *npn* circuit we would expect the forward bias between base and emitter to be such that the base is positive with respect to the emitter. Thus, the positive probe of the voltmeter is placed at the base as shown, with the negative probe at the emitter. The difference between the

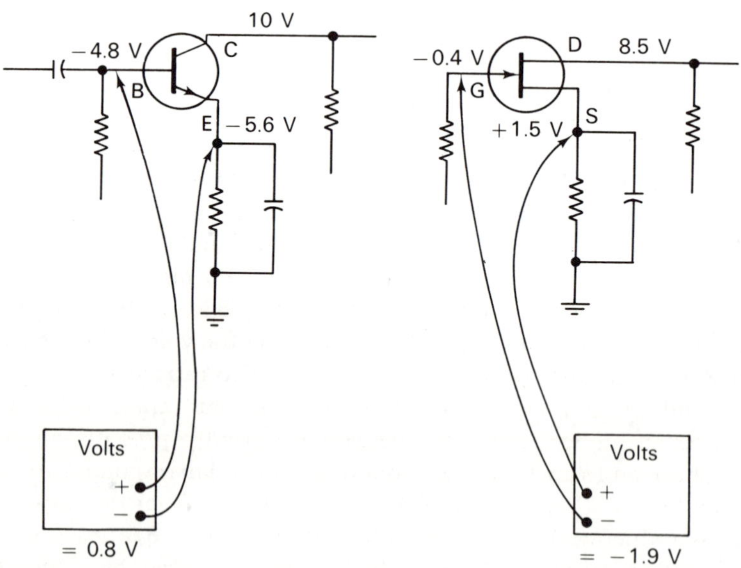

Figure 8-6 Bias check polarities.

—4.8 V at the base and the —5.6 V at the emitter gives us a reading of 0.8 V (positive at the base).

For the FET in Fig. 8-6B similar conditions prevail and the *relationship* between the voltages given must be considered. For the *n*-channel FET the gate is negative with respect to the source element, hence the negative probe of the voltmeter is applied to the gate and the positive probe at the source as shown. The difference between the —0.4 V at the gate and the +1.5 V at the source gives a reading of —1.9 V (negative at the gate input). Thus, for both circuits shown in Fig. 8-6 the bias consists of the difference between the two voltages given, with the polarity being determined by whether a particular element is more negative or more positive than the other element. A reading across the emitter resistor alone (from emitter to ground) in Fig. 8-6A would have indicated —5.6 V and would not have been indicative of the forward bias on the transistor. Similarly, a reading across the source element of the FET shown in Fig. 8-6B would have indicated 1.5 V, and not the true bias voltage.

When reading bias the transistor terminals must be identified for the particular unit in use. Usually the schematic for the device shows the sequence. Typical solid-state terminals are shown in Fig. 8-7 and represent the most common types encountered. Note that similarities occur for transistors and FET units, as well as power transistors and silicon-controlled rectifiers (SCR). Reference should also be made to transistor manuals for listings of the various types. Booklets are also available that list the various types that are interchangeable. For tubes, the number usually appearing at the various elements indicates the socket-hole number. The *bottoms* of the tube sockets are numbered in clockwise rotation, with the first number usually starting beside an open space or at a key notch on the socket.

For years tubes were pretty well standarized with respect to their characteristics and accompanying designation. Hence, a 12AX7 tube, for instance, had the same characteristics and alphanumeric symbol for various manufacturers. This standardization, however, has not been the case with transistors because various manufacturers assigned different alphanumeric symbols to their particular version of a given transistor or FET unit. Consequently, four or five manufacturers may bring out a transistor having virtually identical characteristics but different symbol designations. Thus, various publishers have available replacement transistor manuals so substitutions can be made when a particular transistor is not readily obtainable.

For transistors that can not be identified by schematic reference or that are not marked on the chassis, some basic checks can be made to ascertain which leads are the base, the emitter and the collector. Ohm-

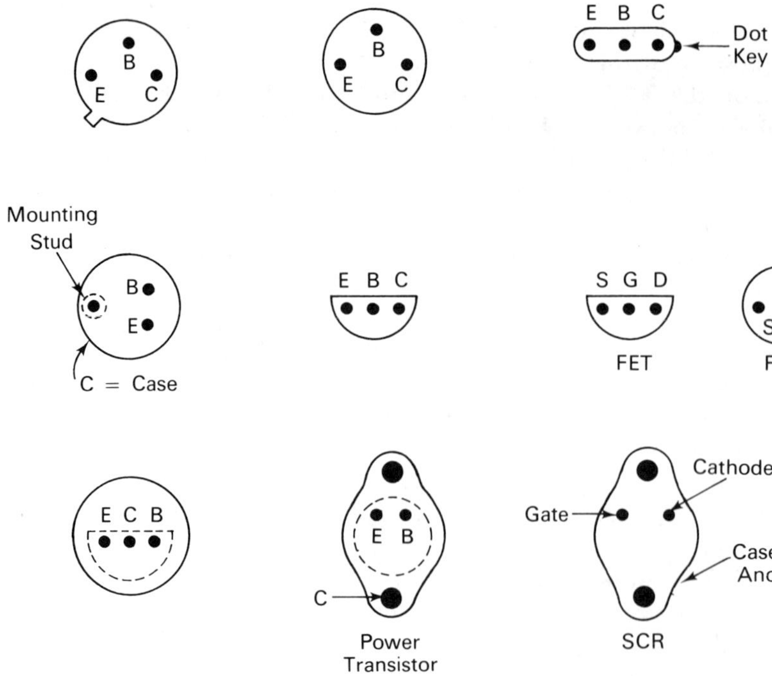

Figure 8-7 Solid-state terminals.

meter checks can be made and commercial testers used to obtain sufficient data for testing, as is more fully described in Sec. 8-4.

8-4 INSTALLATION

When installing new transistors the leads must often be trimmed for proper length. On some early transistors the shock effect of snipping off leads often transferred to the internal solid-state core of the transistor and could damage it. Also, many older transistors were highly susceptible to thermal effects and could be damaged easily during the soldering process. (See also Sec. 8-5 regarding precautions with MOSFET units.)

Even with modern transistors reasonable precautions should be observed if a new one needs to be soldered into place. Heat-sink clamps as shown in Fig. 8-8A are available from electronic parts outlets. These are clamped above the lead to be snipped and soldered, and minimize shock and heat effects. Even with such heat-absorbing clamps, however, the soldering iron should not be applied for too long a period, nor should a high wattage iron be used. A 60 or 100 W soldering iron can

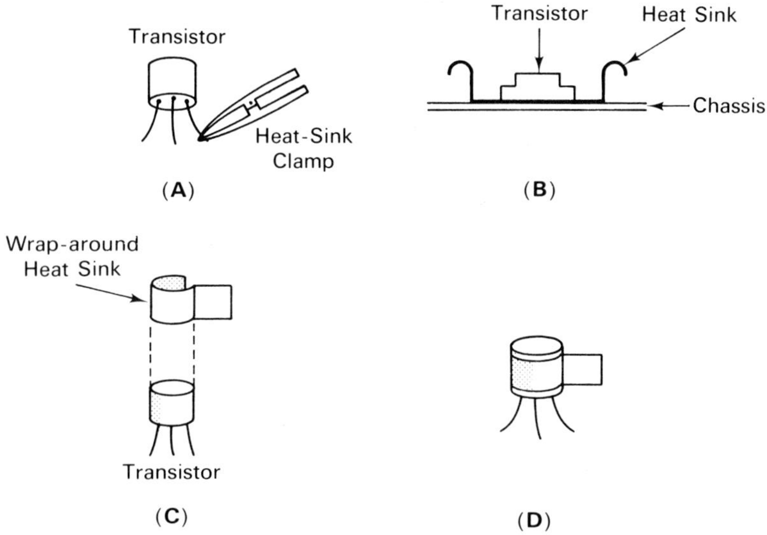

Figure 8-8 Heat-sink devices.

be used, though the parts should be thoroughly clean to reduce the time required to obtain a good soldered joint.

When replacing transistors make sure any heat-sink devices that were present originally are retained. Thus, for the type heat sink shown in Fig. 8-8B the extended flanges are vital to help dissipate the heat generated within the transistor. If the heat sink is left out, the transistor will overheat and burn out. How soon this occurs depends on the wattage being dissipated, the type transistor, and whether circuit faults are present.

Some transistors (such as low-power audio types) utilize a wrap-around heat sink as shown in Fig. 8-8C. This sink unit is pressed over the transistor and appears as shown in Fig. 8-8D. It supplies sufficient heat-dissipating area to prevent resistor damage. Sometimes dual output transistors (such as push-pull types) share a common heat sink, where a single metal strip has each end bent for wrap-around purposes.

For the higher-power transistors of the type shown in Fig. 8-9 special precautions must be observed during installation. Because the outer shell constitutes the collector in the *npn* or *pnp* types it is usually necessary to insulate the shell from the chassis or common ground point. This is done by using a thin mica insulating gasket as shown, pre-perforated to accommodate the mounting bolts and the two prongs representing the emitter and base terminals. A special silicon-base thermal conductive grease is applied to both sides of the gasket to assure maximum heat transfer to the chassis. Thus, in such installations, the metal

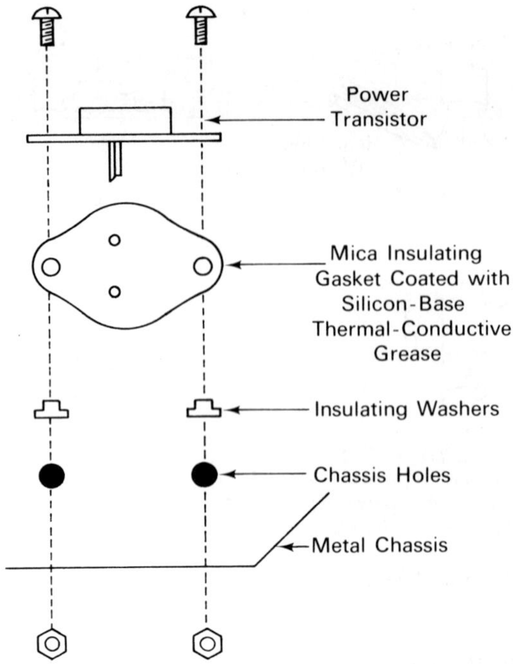

Figure 8-9 Power transistor installation factors.

chassis acts as the heat sink by absorbing the excessive heat of the transistor.

When a defective transistor is removed from a printed-circuit chassis, make sure the correct soldered points are localized on the underside of the chassis. Most of the low-power transistors form a triangular position for the soldered leads as shown in Fig. 8-10. If the transistor leads are bent over after soldering, care must be exercised in removal so as not to break the fragile printed-circuit leads. Since the transistor to be removed is defective, it can be broken if necessary to permit removal of the leads more easily and to protect the printed circuit connections. After the leads have been removed, short-duration applications of the iron may be used to remove excessive solder and to open the holes to accommodate the new transistor.

When installing the new transistor, slip the leads through the printed-circuit connection holes to conform to the sequence of the removed transistor. Cut the excessive lengths from the leads and solder with short applications of the iron. Use rosin-core solder to minimize eventual corrosion of the soldered connections, but make sure sufficient heat is applied to burn away the rosin. Too much rosin deposit around the

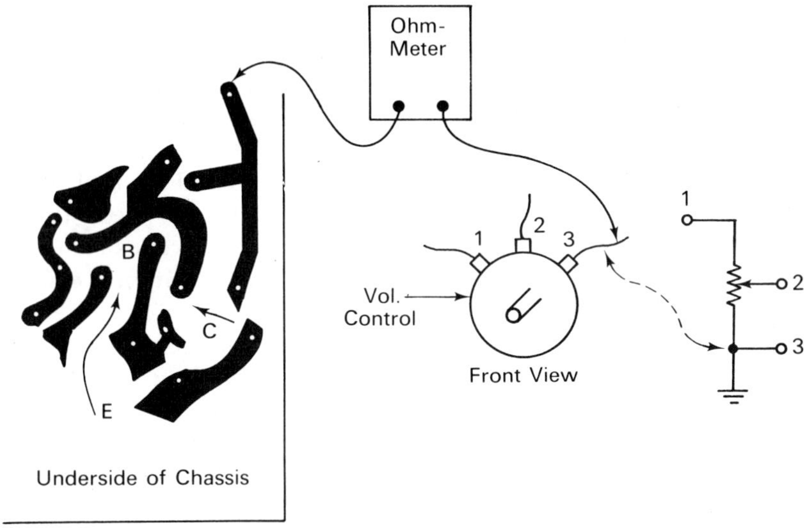

Figure 8-10 Printed-circuit test points.

joint may act as an insulation and prevent the transistor from operating (this is referred to as a cold-soldered joint).

For printed circuit wiring it is difficult on occasion to identify transistor terminals from the bottom of the chassis, particularly if the transistor is situated near the center of the chassis. Some service notes identify printed-circuit leads as well as location of components (see Sec. 9-8). In the absence of such data an ohmmeter must be used to help localize the terminals. If measurements are to be made against chassis ground, the ground lead may often be identified by taking a continuity check from the third terminal of a volume control as shown in Fig. 8-10. For a control having such a ground, of course, there should be full conductivity between the control terminal and that printed-circuit strip representing chassis ground. Any reading above 0 Ω calls for additional checks to locate ground.

8-5 BASIC TEST FACTORS

Some basic ohmmeter checks can be made on tubes and transistors to determine if open or shorted sections exist and hence give a quick indication and localization of the cause of circuit malfunction. For checking gain, emission and other basic characteristics, however, the checking units described in Sec. 8-6 are a necessity.

As shown in Fig. 8-11A, a quick test can be made on a tube to see whether the filaments (heaters) are still intact. If no continuity reading is obtained, the filaments have burned out and the tube must be replaced. (While the bottom-view pins 2 and 7 are often the filament pins, a tube manual must be consulted to make sure other terminals are not used for filaments. Some tubes use pins 3 and 4, others 7 and 8, etc.) If the filaments check out satisfactorily a continuity check should also be made between other terminals of the tube, as well as between the filaments and other terminals to make certain that no shorts exist between tube elements. A tube manual will give details on the internal construction and the isolation of internal leads.

A general continuity check can also be made of transistors by taking ohmmeter readings between terminals. Initially the test leads are applied to two terminals as shown in Fig. 8-11B and the leads are reversed for another reading as at C. There should be a considerable difference in the ohmic values thus obtained, because the reversal of leads has alternately applied forward bias and reverse bias and hence a high conductivity and a low conductivity should occur. If readings are the same (or if no readings are obtained) it is highly probable the transistor is defective.

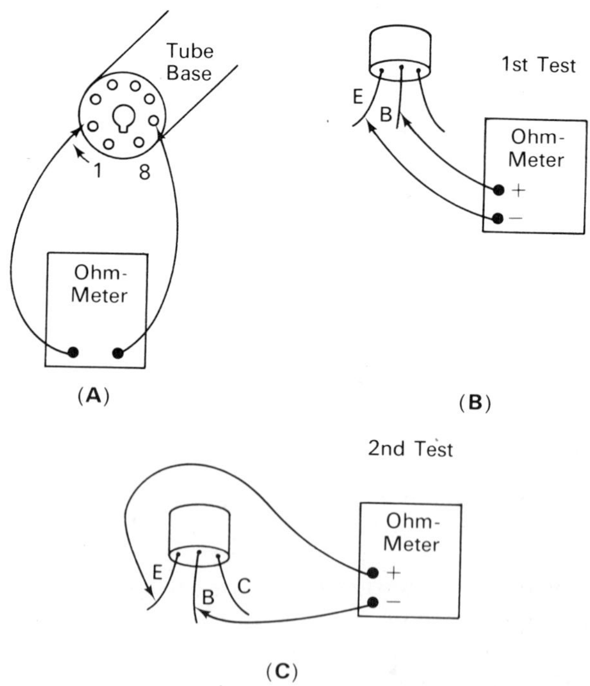

Figure 8-11 Basic conduction tests.

When an old transistor is removed and replaced with a new type an identical installation can be made by careful observation of the leads. Hence, it is not actually essential that the technician be able to identify the collector, emitter or base. Also, for the tests shown in Fig. 8-11B and C we need not know the actual sequence of emitter, base and collector because we are simply ascertaining if the proper high-ohm and low-ohm readings are obtained when we apply the leads to any two terminals and reverse such leads. Actually, however, if the transistor in Fig. 8-11B is a *pnp* type a high-ohm reading will be obtained provided the ohmmeter test leads have been properly identified regarding internal battery polarity. (This can be ascertained by taking a voltmeter reading of the ohmmeter leads by using another meter.)

For the polarity shown in Fig. 8-11C a low-ohm reading is obtained. If the transistor is an *npn* type the test connections at B will give low-ohm readings and those at C will give high-ohm readings. (High- and low-ohm difference will, of course, also be obtained for continuity tests between base and collector.) When more information is required than simple continuity tests, the commercial checkers should be used for they provide for lead identification, gain factors and other performance data.

For continuity checks on the field-effect transistors special considerations prevail and differences exist between the JFET and the MOSFET units. If the forward resistance of a JFET is to be read, an ohmmeter should be used that has a low-voltage source. This low-potential condition is found in the $R \times 100$ range of many meters, particularly those having 50 or 100 μA meter movements.

When the JFET unit is an *n*-channel type, for forward resistance the negative lead of the ohmmeter is applied to the gate terminal and the positive probe to the drain or source terminal. (Again, the correct polarity of the ohmmeter leads must be established because the positive probe socket in a VOM will refer to voltmeter and ammeter readings, and not necessarily the ohmmeter polarities. As mentioned, use a separate meter to read voltage polarities at the terminals.) For *p*-channel testing, of course, the leads are reversed for reading forward resistance.

For checking reverse resistance of the *n*-channel type JFET, the positive lead of the ohmmeter is applied to the gate and the negative probe to the source or drain. If a JFET is not defective the reverse-resistance test should show an almost infinite resistance reading. For the *p*-channel reverse resistance, the negative lead is connected to the gate and the positive to the drain or source.

Many MOSFET units have very critical and sensitive internal characteristics and can be damaged by improper handling and testing (unless it is a gate-protected MOSFET). Unusual precautions are often stipulated for certain MOSFETs, such as shorting leads together by metal spring connectors when handling the unit. Similarly, when applying a soldering iron the tip should be at ground potential (heated by the iron's element but isolated from it electrically).

The forward resistance (and reverse resistance) of a MOSFET can be tested with a low-voltage ohmmeter provided the probes do not apply a voltage that is sufficient to start transistor function. Many modern VOMs have a low-power mode for in-circuit ohmmeter checks but use a sufficiently low voltage to prevent transistor conduction. Some such meters use only 0.08 V, while in the regular mode of operation 1.5 V or more may be used.

For the MOSFET, a resistance reading taken from the gate to drain (or from gate to source) should indicate a virtual infinite resistance for both forward and reverse resistance checks. As with junction transistors and tubes, such tests indicate only shorted sections, but will not indicate a change in characteristics that affect gain and other operational characteristics. For a thorough test of the FET, as with junction and other transistors, modern commercial testers should be used as described in the next section.

8-6 COMMERCIAL TESTERS

Commercial transistor and tube testers come in a variety of models; generally their versatility depends on a number of built-in features, all of which add to the cost of the instrument. Separate transistor testers are available, as are standard tube testers and individual testers for cathode-ray tubes only (primarily for picture tubes). Some hybrid models are available that provide for transistor and tube testing in a single instrument.

The most common tube tester is the *emission* type wherein fixed voltages are applied to the grid and anode (with respect to the cathode) and the amount of tube conduction is then measured. Because voltages are constant, this is known as a *static* test. Generally a simple dial face is used, showing whether the tube is defective, in the questionable area (indicated sometimes with a question mark) or good, as shown in the upper portion of the dial face for the transistor tester of Fig. 8-12.

For precision laboratory work a higher quality (and higher priced) tester is used that measures the characteristics of the tube under changing signal conditions. Such a test is a *dynamic* one, hence checks the tube for its characteristics under actual signal-handling conditions. The characteristics tested with laboratory instruments is the *transconductance,* which is the ratio of plate-current change to a grid-voltage change, with

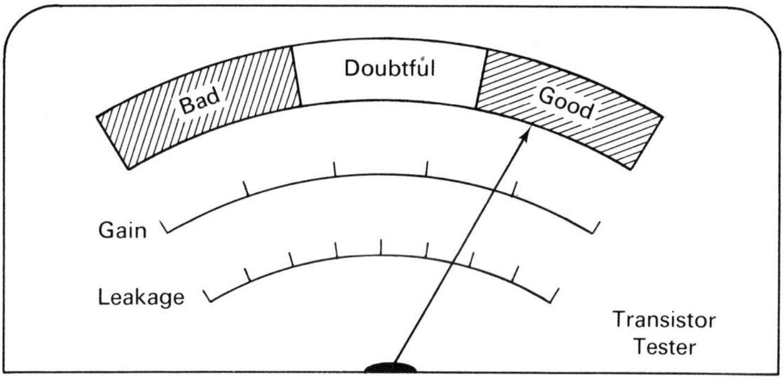

Figure 8-12 Typical transistor tester dial.

the plate voltage held constant. The equation is given here for reference, with *d* representing a *change* of value:

$$\text{transconductance } (g_m) = \frac{dI_p}{dE_g}\bigg|\, E_p \text{ constant} \tag{8-1}$$

Thus, transconductance is an approximate figure of merit for a tube and shows how much signal current change is created for a certain amount of grid voltage change. This characteristic is also referred to as *mutual conductance* and the unit quantity is *mho* (ohm spelled backwards).

Transconductance also refers to the FET and is expressed as:

$$g_m = \frac{dI_d}{dE_g} \tag{8-2}$$

where dI_d = small change in drain current
dE_g = small change in gate voltage

For the general run of transistors, the characteristic referred to as the forward current-transfer ratio takes the ratio of a change in collector current (I_c) to base current (I_b). This is known as *beta* and uses for its symbol the Greek lower-case letter of the same name:

$$\beta = \frac{dI_c}{dI_b} \tag{8-3}$$

For the higher-quality commercial transistor testers a beta range is also given as shown in Fig. 8-12, with the gain represented in micromhos. Dual scales are sometimes provided with ranges up to 500 μmhos for one scale, and another giving still higher ranges for versatility in testing a variety of transistors.

The leakage test is useful for determining mA or μA of current leakage (usually between the base and collector junction under reverse current conditions). This is an important test since even 1 or 2 μA of leakage can cause poor performance in many transistors. Such testers also provide for leakage checks in FET units between gate and source elements.

When using commercial checkers transistor data are furnished by the manufacturer so that the tester can be adjusted properly for the particular transistor to be read. Some commercial meters provide for in-circuit tests, eliminating the need for unsoldering leads. Other testers also have provisions for identifying emitter, base and collector leads and ascertaining whether the transistor is a *pnp* or an *npn* type. Thus, in the purchase of such testers a comparison should be made of the variety of tests each will perform in relation to the purchase price and to the requirements of the technician making the purchase.

Picture-tube testers are also available in various models and with a wide range of capabilities. A typical front-panel layout is shown in Fig. 8-13. In this unit other tests are available in addition to the standard emission test that gives a general indication whether the tube is bad, questionable or good. The tracking scale permits checking of the emission level of the individual red, green and blue cathode and gun assemblies. Since proper settings of filament and application of correct bias potentials are necessary, a tester that makes the process automatic (without having to calculate current changes and relationships) is a desirable type.

8-7 ICS AND MODULES

Modern manufacturing procedures for transistors and other solid-state devices utilize electronic and chemical processes to reduce electronic circuitry to an extraordinary degree by the process known as *microminiaturization*. Thus, a single silicon chip or a so-called *integrated circuit* (IC) can contain dozens of units including diodes, transistors, FET's, resistors and associated wiring. Such ICs are often limited in size only by the necessity for bringing out connecting leads. Once such IC units have processed signals with their internal circuitry, the signals are amplified an additional amount and applied to speakers, recorder heads or other units as needed. For low-power audio output, however, the final stages can also be contained within the integrated circuit.

Printed circuits are also important to the microminiaturization of electronic circuits because they eliminate bulky interconnecting wires

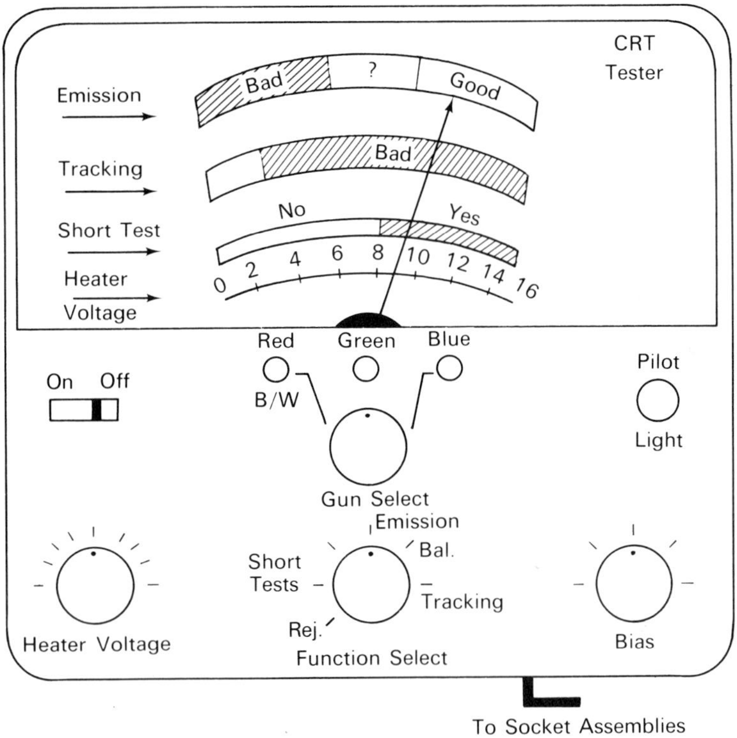

Figure 8-13 Pix tube tester.

and reduce undesired effects of stray capacitance between hook-up wires. Thus, they also lend themselves to mass-production techniques. Also, since hook-up wire lengths are reduced, there are decreased effects of undesired inductance contributed by the wire lengths—particularly at the higher frequencies.

Integrated circuits have many shapes, with some square, others circular and still others rectangular. The IC's have a number of prongs designed to fit into sockets or into holes in the printed-circuit board for soldering to the metallic conducting strips.

For the circular IC shown in Fig. 8-14A, a base extension (or notch) identifies the start of the identifying numbering system. For the bottom view shown, the sequence is clockwise. For the rectangular IC shown in Fig. 8-14B, the connecting segments are numbered also in consecutive order with a notch providing identification of the start of the numbering. On occasion an imprinted dot may be used to show the beginning of the count sequence. The side view of the rectangular IC is shown in Fig. 8-14C, plus the socket often used for convenience in replacement.

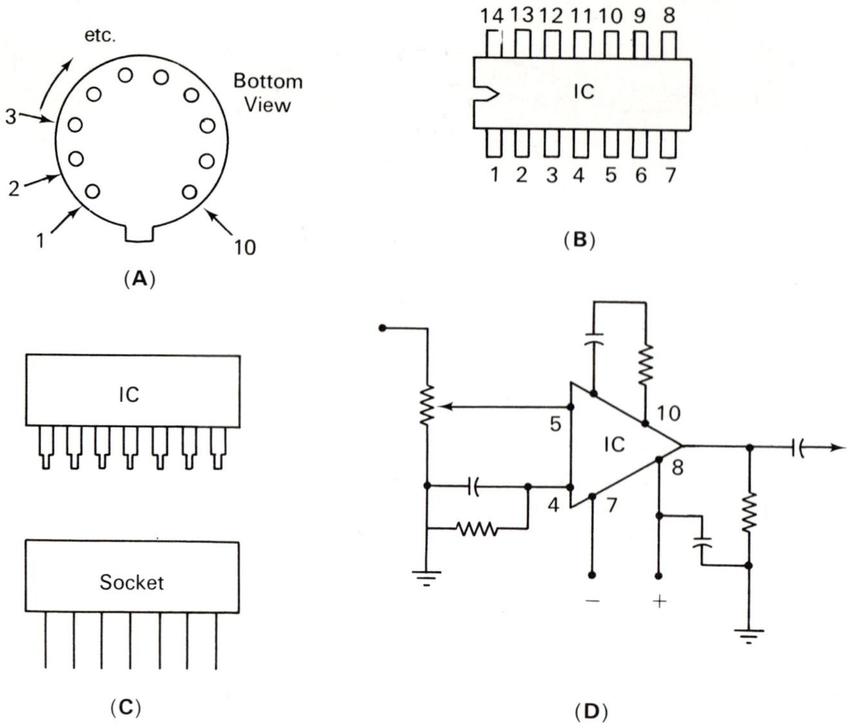

Figure 8-14 Typical ICs.

While the socket increases the bulk of the assembly, it eliminates the necessity for unsoldering prongs when a defective IC is found.

Some of these units measure less than a half inch in length, depending on the number of circuits within them and the necessity for furnishing good contacts for soldering purposes.

IC units are often illustrated schematically in triagular form as shown in Fig. 8-14D, though square and rectangular representations are also common. Only those terminals actively used are numbered in the schematic. If more than one IC is used, the usual procedure is to identify them as IC_1, IC_2 and IC_3.

When modules are used each may contain several ICs to make up a particular circuit system such as an IF and detector system or detector and output amplifier sections. Many modules can be installed by pressing the unit into place, since appropriate prong and prong receptacles are present for convenience in replacement. Of course, separate ICs may be used for these functions as shown in Fig. 8-15, where one IC constitutes the circuitry associated with the sound IF stages, and the second IC combines the sound detector with the sound output amplifier.

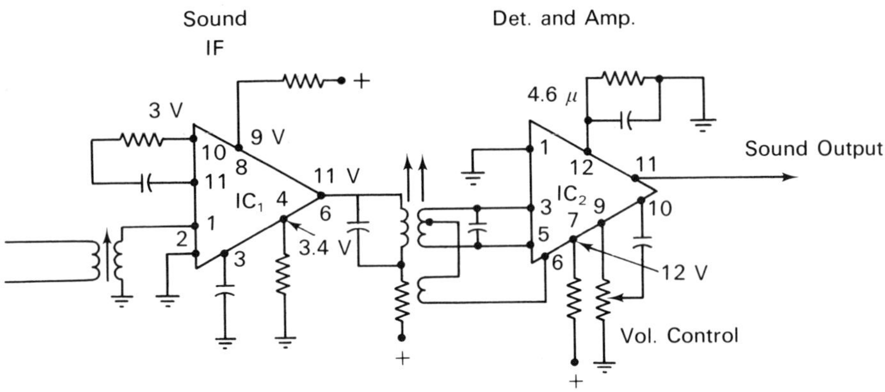

Figure 8-15 IC voltage points.

In many schematics using ICs, various voltages present at the terminals are given as exemplified in Fig. 8-15. Thus, measuring from terminal 4 of IC_1 to ground, we would expect to read + 3.4 V and any deviation from this value would call for additional tests. Similarly, 3 V should be read at terminal 10 and 9 V at terminal 8. At the output we should obtain 11 V (terminal 6). Similar voltage designations are shown for IC_2 to aid in troubleshooting.

The circuits illustrated in Fig. 8-15 are those that are present in either an FM receiver or in the FM sound section of a television receiver. If voltages check out satisfactorily, but sound output levels are down, the interstage transformers may need tuning. If tuning touch-up fails to correct the problem and the speaker system checks out for open circuits or shorts, new ICs may have to be tried. Signal tracing procedures with a scope will ascertain whether there is any output from IC_1. If IC_1 does not show any output the unit may be defective. If an output is obtained from IC_1, but not from IC_2, then of course IC_2 must be replaced. Replacement should not be attempted, however, unless all other factors check out satisfactorily, including properly tuned interstage transformers, and correct voltages at each point designated.

9

stereo FM and hi-fi testing and troubleshooting

9-1 GENERAL SERVICING FACTORS

Stereo and quadraphonic systems, despite the greater amount of circuits involved, are often easier to test and troubleshoot than mono (single channel) systems. If, for instance, one channel of a stereo or quadraphonic systems fails to operate, we can make comparisons of voltage readings and scope tests between the defective channel and the one that is operating satisfactorily. Thus, differences in readings will provide clues to identify the defective component.

Servicing procedures have already been covered in earlier chapters on various sections of stereo and high-fidelity equipment and reference should be made to such sections as apply, particularly Sections 6-4, 7-1, 7-2, 7-4, 7-5, 8-2 and earlier sections that relate to basic considerations regarding leaks, partial or complete short circuits, continuity testing, distortion checks and related subjects. (See also Figures 1-6, 4-18 and 6-5.)

To expedite testing and troubleshooting the technician should be familiar with the general characteristics of stereo and hi-fi systems as covered in the early sections of this chapter. A basic understanding of the system and the characteristics and purposes of specific circuitry simplifies localization of defective parts because logical reasoning can be applied to interpretation of symptoms.

9-2 STEREO FM MULTIPLEX SYSTEM

Multiplexing is a system whereby several signal types are transmitted simultaneously. This process is necessary in FM stereo broadcasting to accommodate the additional sidebands required when more than one channel of audio is involved. An additional carrier (called a *subcarrier*) is used for the modulation process that produces the required sidebands. Once the sidebands have been obtained, the subcarrier is suppressed at the transmitter to save spectrum space. Hence, at the receiver, the subcarrier must be developed in the receiver and recombined with the sidebands before the detection process can be performed as described more fully later in this chapter.

The basic frequencies involved are shown in Fig. 9-1. The subcarrier's frequency is 38 kHz. Because the receiver must produce this signal accurately it must receive a signal that can be used for synchronization purposes. Such a synchronization signal consists of a 19 kHz *pilot* subcarrier. This pilot signal is doubled in the receiver to the required 38 kHz frequency as shown later.

As shown in Fig. 9-1, the complete multiplex modulation signal for stereo FM consists of a mono L + R signal in the audio range (50 Hz to 15 kHz), a supersonic 19-kHz pilot subcarrier signal, plus the L − R

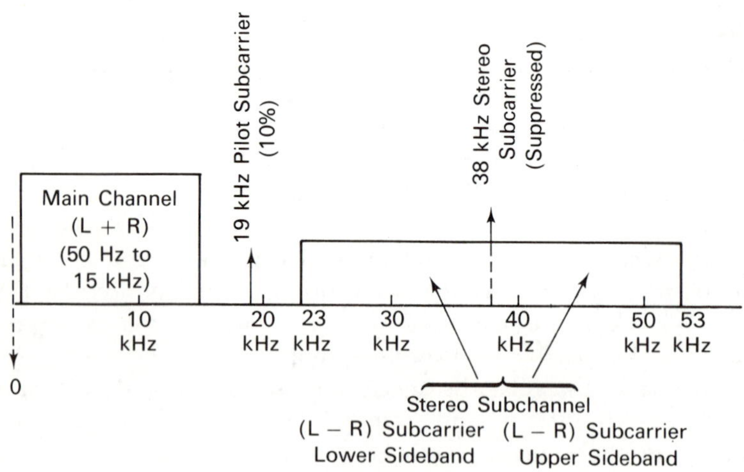

Figure 9-1 Stereo FM and SCA spectrum.

supersonic signal (23 to 53 kHz) with suppressed 38 kHz carrier. As with the mono FM signals, the audio-signal frequency range is at least 50 Hz to 15 kHz for a high-fidelity audio range.

9-3 STEREO FM RECEIVER SYSTEM

As with AM and FM radios, the stereo FM radio uses the superheterodyne principle illustrated earlier in Fig. 6-3. Instead of the single diode detector used in AM or TV, however, dual diodes are employed as shown in Fig. 6-5. In mono FM the detector output incorporates a series resistor and shunt capacitor to bypass a portion of the higher-frequency signals that were emphasized at the transmitter. This is done to lift the signals above the noise level. The *pre-emphasis* of the higher-frequency signals improves the signal-to-noise ratio in FM. Thus, a de-emphasis network is used at the receiver to bring the response of the audio spectrum back to normal (circuitry is shown later in this chapter).

A block diagram of the multiplex (stereo) FM receiver system is shown in Fig. 9-2. Note that the initial block (to which the antenna is attached) consists of a complete FM mono receiver containing the tuner circuits, the intermediate-frequency (IF) stages and the ratio detector. The de-emphasis circuitry is omitted here, however, and used later in the added circuits as illustrated.

The output from the ratio detector contains all the signal types received. If a stereo signal is received, the output from the ratio detector contains the left-channel and right-channel sidebands, plus the pilot subcarrier, and these are accompanied by mono signals to form what is termed the composite FM signal. As shown in Fig. 9-2, a special amplifier raises the composite-signal amplitude to that required for processing.

The output from the composite-signal amplifier is applied to both a 19 kHz amplifier and a 19 kHz filter circuit. From the 19 kHz amplifier a doubling process is used as illustrated later and the output is applied to a balanced-bridge rectifier system. The 19 kHz bandstop filter passes all signals except the 19 kHz pilot carrier; hence its output contains the 50 Hz to 15 kHz demodulated L + R signals, the 23- to 53-kHz L − R sideband signals and a 67 kHz SCA signal, if such is being transmitted by the particular station being received. The SCA signals are sent when the station provides background music to stores and business offices. A decoder is needed to receive such special broadcasts. Thus, for stereo FM for homes, the SCA signals are removed by the 67 kHz suppression filter. The remaining signals are then applied to the balanced-bridge demodulators in conjunction with the 38 kHz subcarrier signal.

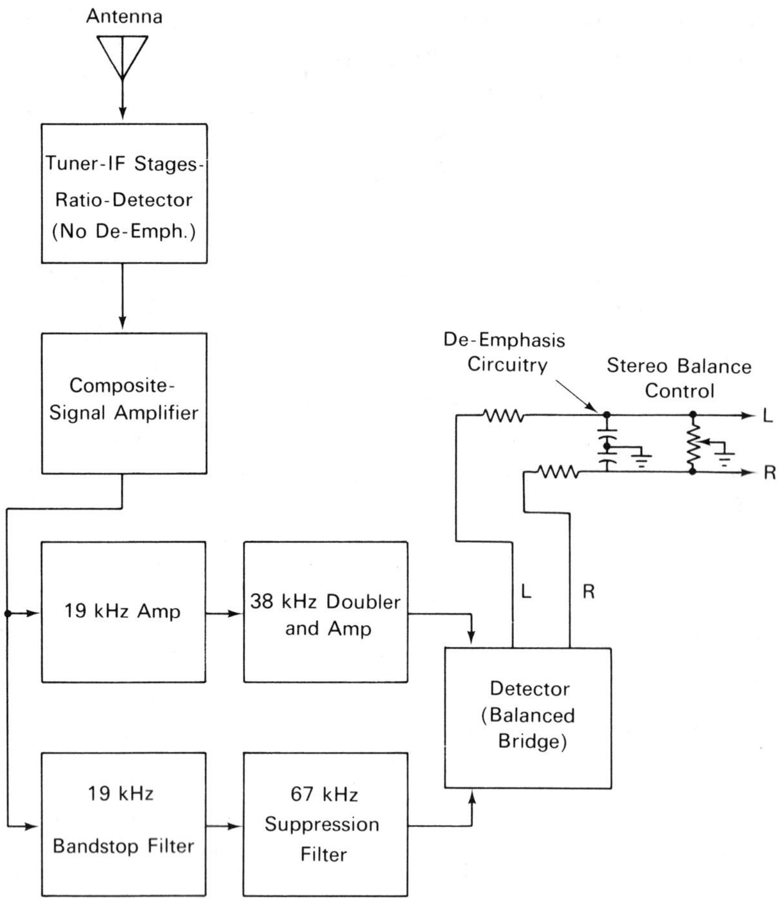

Figure 9-2 Multiplex (stereo) FM receiver system.

Within the demodulator (detector) circuit, the sidebands are mixed with the subcarrier and detected. The left and right channel signals are individually de-emphasized as shown, and a common potentiometer across the two output lines provides for adjusting the balance between the L and R channels.

9-4 MULTIPLEX RECEIVER CIRCUITRY

The circuits between the 19-kHz amplifier and the stereo detectors are shown in Fig. 9-3. Transistor Q_1 amplifies the pilot subcarrier signal and applies it to transformer T_1, where C_4 forms a resonant circuit with the primary to T_1, thus tending to accept the 19-kHz signal and reject others in typical resonant-circuit selectivity fashion. Note that transistor Q_2 (the 38-kHz amplifier) has a positive potential applied to the base (using resistors R_6 and R_7) from the + 20-V source. Thus, this transistor is biased in reverse at the base-emitter input and does not conduct unless such reverse bias is changed to the forward-bias condition.

On the collector side of Q_2 is resistor R_9 in series to ground. If the transistor does not conduct, no current flows through R_9 and no voltage drop occurs across it. Thus, no forward bias is developed for the indicator-control transistor Q_3 (which is an *npn* type in this instance and needs a positive potential at the base with respect to the emitter). If Q_3 is unable to conduct, the stereo indicator lamp (in series with the col-

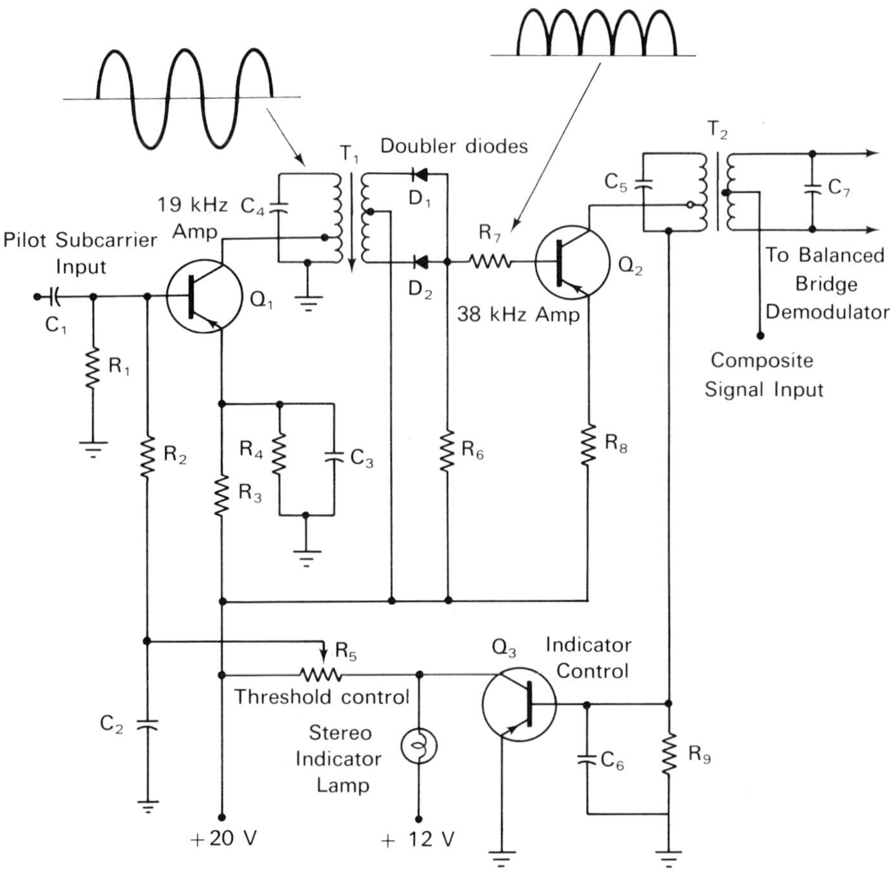

Figure 9-3 Stereo FM pilot amplifier and doubler.

lector feed line of Q_3) has no current flow through it and hence does not light up. Thus, mono signal reception is indicated, or the tuning is between stations.

When a stereo signal is tuned in, the 19-kHz signal is applied to the input of Q_1 and amplified. The signal across the secondary of T_1 is applied to the full-wave rectifier-circuit diodes D_1 and D_2. These diodes form a doubler circuit that produces successive alternations at a repetition rate twice that of the pilot carrier. These pulse a 38-kHz resonant circuit (C_5 and the primary of T_2) and the flywheel effect produces a sinewave-type 38-kHz carrier for application to the balanced-bridge demodulators.

Successive full-wave alternations switch transistor Q_2 into periodic conduction at the 38-kHz rate. During conduction the collector current flow through resistor R_9 produces a voltage drop that is positive at the top of the resistor (and hence at the base of Q_3). The positive potential at the base of Q_3 supplies the required forward bias and Q_3 conducts, permitting the stereo indicator lamp to glow. Capacitors C_2 and C_6 filter signal components from these sections to produce the required steady-state dc bias.

The circuitry shown in Fig. 9-4 is a continuation of that illustrated in Fig. 9-3. The composite signal, minus the 19- and 67-kHz signals, is applied to the center tap of the secondary of T_2. The four diodes constitute a balanced-bridge system for processing the incoming 38-kHz signals in relation to the composite signals with sidebands. The composite signal at the center tap appears in phase at any instant of time at the top and bottom of the bridge rectifier system. The incoming 38-kHz carrier, however, places out-of-phase signals across the bridge for any instant of time. Changes cause greater amplitude of detected signals when in-

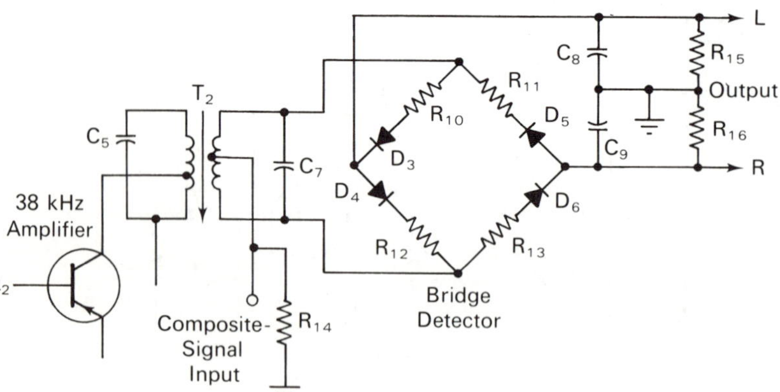

Figure 9-4 Stereo detector.

phase conditions occur, and lower-amplitude outputs for out-of-phase conditions. Capacitors C_8 and C_9 act as filters and change the rectified pulses into an average-value voltage that varies in frequency and amplitude to conform to the audio-modulating component that had been contained on the original left and right channels.

As shown in Fig. 9-4, the output audio signals are developed across R_{15} and R_{16} as against common ground center tap. Signals are then fed to standard audio amplifiers as shown later.

9-5 MULTIPLEX ALIGNMENT FACTORS

The resonant circuits of transformers T_1 and T_2 shown in Fig. 9-3 have variable-core tuning. Alignment procedures consist of tuning these transformers for maximum signal output. Some touch-up procedures can be undertaken by tuning in a stereo station and making slight adjustments if needed. It must be emphasized, however, that no alignment should be undertaken unless one is sure that the problems are actually caused by misalignment. If the receiver has been working well and suddenly develops poor reception problems, it is unlikely that alignment is necessary. Thus, adjustments could actually throw the receiver into misalignment. If touch-up procedures are undertaken, try to ascertain the original settings of the adjustable slugs so as to be able to return them to this position if necessary.

For proper alignment a multiplex generator is recommended. Commercial models provide an output consisting of a composite multiplex signal as well as an RF signal that is frequency-modulated by the multiplex signals. This RF signal is injected into the antenna terminals and the 19-kHz transformer is adjusted initially for maximum output as indicated on an oscilloscope reading at the transformer output. Next, the 38-kHz transformer is tuned for maximum output. If the multiplex generator has meters for determining stereo separation, switch off the modulation for one of the channels and readjust the T_1 and T_2 transformer cores for maximum separation as shown on the meter readings.

The multiplex generator input at the antenna terminals is a more reliable procedure than injecting the signal at the output circuit of the FM ratio detectors. Phase shifting of signals in the detector circuitry will contribute to inferior results if the signal is injected there.

Because the front section of a stereo receiver is identical to standard mono FM receivers, tracking and alignment of the tuner and IF stages is the same as that detailed in Chapter 7. (see Sections 7-1 and 7-5). Also, signal-tracing procedures and other test factors as detailed for other circuitry in earlier chapters also apply to multiplex stereo receivers.

When troubles arise in stereo reception, however, the multiplex circuitry must be checked. If, for instance, mono reception is obtained even when tuning to a stereo station, defective transistors Q_1 or Q_2 may be the cause, or circuit components may be defective in the circuitry of Fig. 9-3 and 9-4.

Distortion of the audio signals in one (or both channels is usually caused by circuit faults in the audio-amplifier stages and these should be checked for faulty transistors or defective components according to the procedures outlined in earlier chapters. Additional troubleshooting data and tests are given in the remaining sections of this chapter.

A stereo receiver (as with a radio or television set) must not be evaluated on the basis of reception from a single station. A particular station may be broadcasting music from a mono record, even though they are transmitting in a stereo mode. Also, a particular station may be having transmission difficulties at any given time, or reception may be poor because of an inadequate antenna system or because the station is too distant from the receiver. Always check stereo reception by listening to several stations, preferably local ones. If the stereo indicator light does not identify stereo transmission, check for a burned-out indicator bulb or a defective indicator-control transistor. For Fig. 9-3 there should be a voltage change across R_9 when a stereo station is tuned in. Also, check for proper adjustment of the threshold control potentiometer. (Distant stations that put in a weak signal will not provide for sufficient amplitude to give a bright indication on the stereo lamp. In most instances, however, such weak stations also would be received poorly because of decreased signal to noise ratio. If reception for a particular weak station is desired, it is best to switch the receiver to the mono mode for better signal-to-noise characteristics. For a given broadcast of stereo, the switch to mono will provide for a noticeable noise decrease for the weaker stations.)

If better reception of distant stations is desired on a consistent basis, a better antenna system (preferably outside) must be considered. (See Sec. 1-4 for additional data.) Couplers can also be obtained so that a television antenna can be used to feed both the TV set and an FM receiver (see Sec. 5-8).

9-6 AM-FM STEREO RECEIVER TESTS

Stereo FM receivers are available with an audio power output ranging from a few watts per channel to well over 150 W per channel. Some are dual receivers and have provisions for switching to either AM audio or FM radio. The more costly types have added features including remote-speaker switching provisions, additional inputs for tape recording monitoring, various push-button controls for low- and high-frequency response adjustments, as well as special noise controls. Most have balance controls for adjusting the relative gain of left and right channels, plus a single-knob volume control (using ganged potentiometers).

Lower-priced models often do not accept magnetic phonograph cartridge input because no preamplifier provisions are included. Thus, either a crystal or ceramic cartridge must be used for adequate volume output from phonograph playback. Crystal or ceramic cartridges have a greater signal voltage output than do the magnetic types, and hence will overload circuitry if injected into a plug designed for *magnetic* cartridges. The more elaborate receivers have inputs for both the magnetic or crystal.

The basic switching system for an AM-FM receiver is shown in Fig. 9-5. In addition to the two inputs for phono cartridges, this receiver has an auxiliary input for playing tape decks such as cassette, cartridge or reel-to-reel types. A tape output jack is also provided to pick up signals after the first or second stage of audio amplification (the x and y points shown in Fig. 9-5). This jack permits the recording of any input provided by the selector switchs (ganged for single control operation).

As with most receivers of this type the output speaker circuitry has provisions for listening with stereophonic earphones. A series 330 Ω resistor is in each earphone line as shown, and the speakers are disengaged when the phones are plugged in. With high-power audio systems it is extremely important that no shorts occur across the speaker output terminals to prevent output-transistor burnout. Best practice is to have the speakers connected before turning the receiver on. Well-insulated speaker cord should be used (or good quality lamp cord, with #18 wire).

For proper stereo reception the loudspeakers must be properly phased, as shown in Fig. 9-6. If the right terminal of one speaker connects to the lower output terminal of the speaker panel as shown, the second speaker's right terminal should also be wired to the lower output terminal, as seen in Fig. 9-6A. Also, the left speaker terminals connect to the top output terminals. If the speakers are not properly phased it will result in a less pleasing sound and improper tone source characteristics. Phasing can be tested by switching to the mono mode (or playing a mono record) with the balance control set at center. Now the sound source should appear to come from between the two speakers as shown in Fig. 9-6B. (The same effect should be apparent when listening with stereo earphones.) If phasing appears incorrect, reverse the wires to *one* (not both) speakers and recheck results.

With a quadraphonic receiver system, speaker phasing procedures become more difficult if the speaker output terminals are not properly

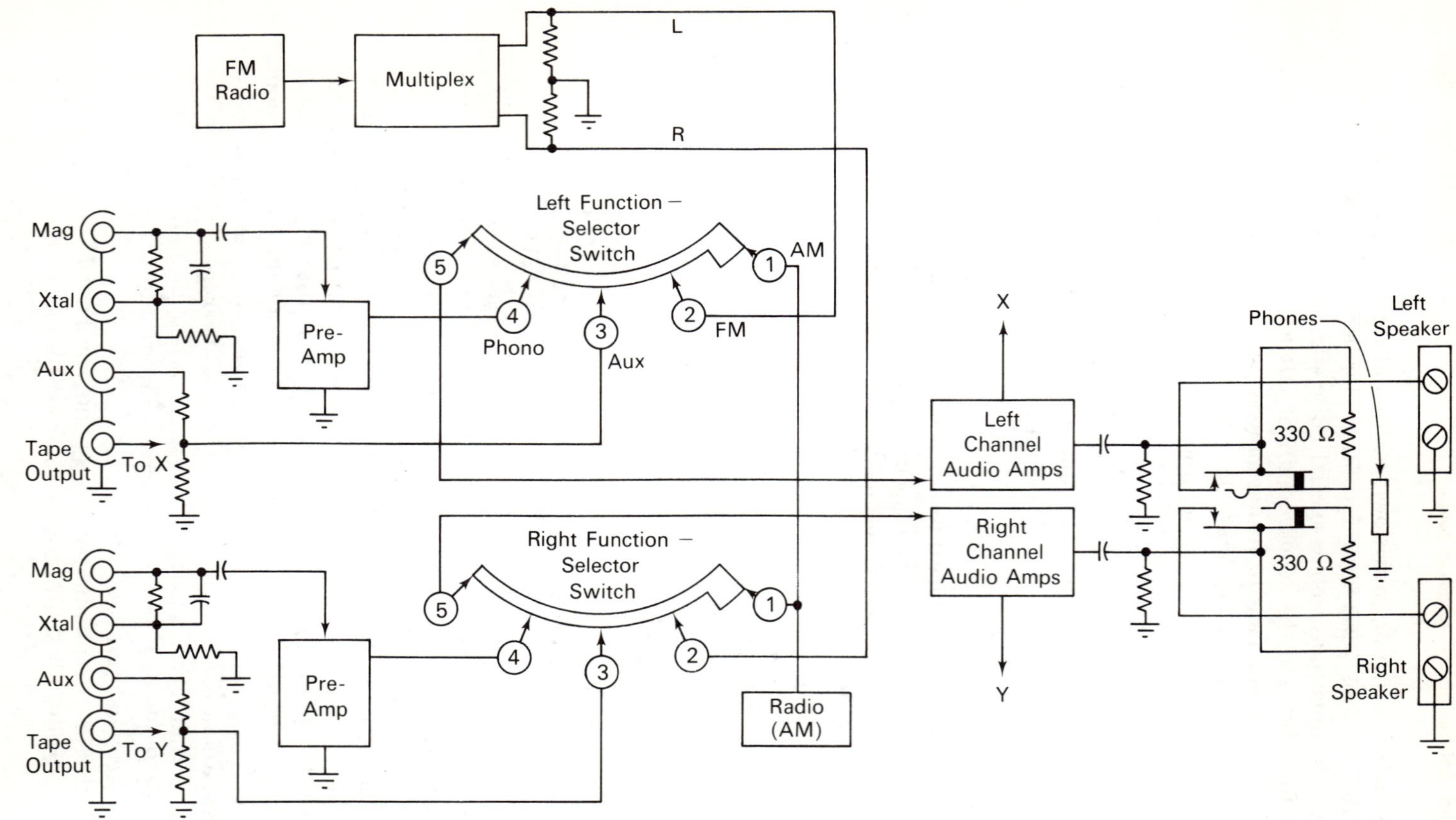

Figure 9-5 AM-FM stereo radio switching system.

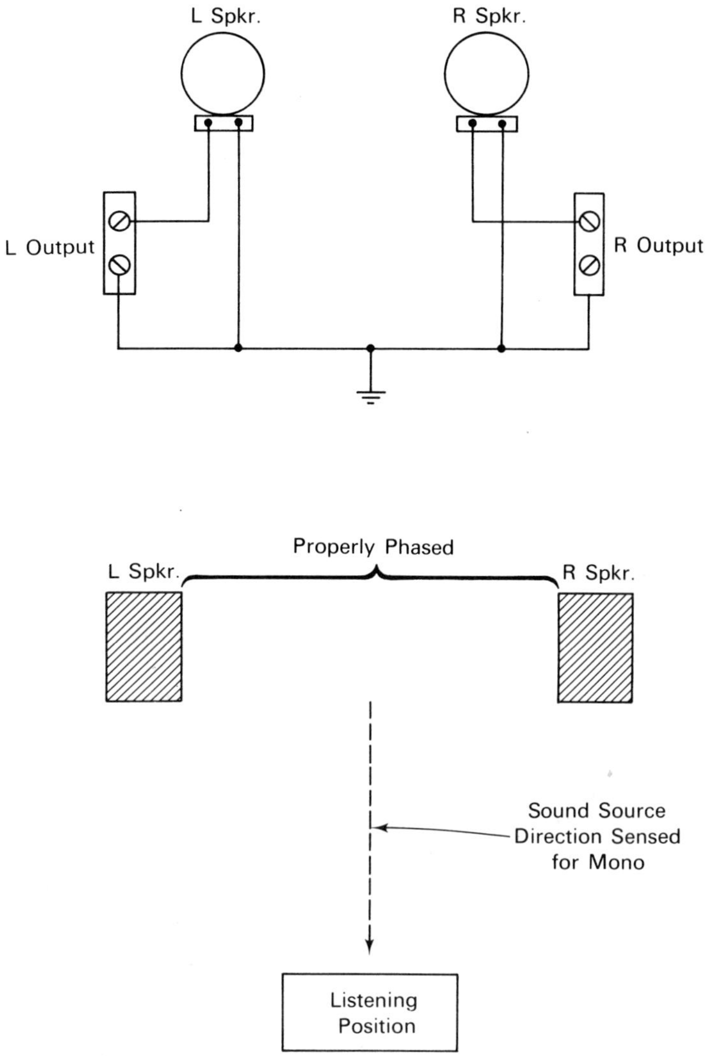

Figure 9-6 Stereo-speaker phasing factors.

marked for the common ground links and the speaker leads are not identified for the respective connections of each wire to the speakers internal connections. Thus, the backs may have to be removed from speakers to identify lead connections. For a quadraphonic system, phasing can be checked for two speakers at a time by turning down the gain of the individual amplifiers if such adjustments are provided, or by unplugging two input lines. A mono record again can be used to check for the presence of the center sound source (between any two speakers).

Some quadraphonic systems and records do not have the distinctive separation achieved with stereo, hence overall evaluation of the end results is difficult unless the performance for a particular system is compared to another using the same recorded material. Test records are available for both stereo and quadraphonic systems and are valuable in servicing and set-up procedures because they provide for a variety of tests, including channel separation, signal-to-noise ratio, frequency response, harmonic distortion, channel sensitivity and so forth (see Sec. 9-9 and Sec. 9-10).

When troubles occur in receiver systems, some basic checks can be made without using test instruments. The earphones, for instance, can be used to check for the cause of unequal speaker response. Also, speakers can be interchanged as a quick check if performance appears inferior in one channel as compared to the other.

For instance, if there is no sound output from the right speaker, it can be plugged into the left output and the left speaker into the right output. If the speaker is defective no sound will appear from the left channel now, but the right sound output will be present. In lieu of this procedure, the earphones could be plugged in to ascertain whether each channel is providing for sound output. If each channel is heard equally in the phones, check for open wiring in the dead speaker system, particularly in the stereo jacks and plugs. Similarly, if one earphone gives no sound (or intermittent sound), but speaker reception is normal, look for breaks in the plugs and cables to the phones. Usually a new three-wire earphone cable solves the problem. (The factors discussed in Chapter 1 should always be kept in mind, for trouble conditions are usually caused by open circuits, short circuits or partial shorts.)

If one channel is dead (or has very low volume) and the speaker system checks out satisfactorily, the wiring of the input accessories should be checked for open or short circuits. Again, a quick check can be made by changing the input lines. If, for instance, a phono is being used, disconnect both input plugs then insert one plug only into one channel to check sound output. If sound is heard, put the same plug into the other channel. Repeat the single plug-in procedure for the other plug, thus isolating a possibly defective phone line. The same tests can be made with the input lines from tape players or other inputs when used. If the input lines test out, checks will have to be made of the circuitry (signal tracing, continuity checks, component testing and voltage readings as discussed in earlier chapters).

9-7 TAPE DECK SERVICING

Tape decks or tape players of all types have a basic switching system as shown in Fig. 9-7. If the system is stereo, everything except the bias oscillator will be in duplicate, with common switching and control functions. A *deck* is designed to be used with an existing amplifier system (either a separate amplifier or the audio amplifier that is part of the radio). The deck does not have audio power amplification for operation of speakers, though sometimes sufficient audio power is present for earphone monitoring purposes. The tape *player* is provided with increased audio amplification to bring the sound level to the amplitude required for loudspeaker operation.

As shown in Fig. 9-7, a microphone input is provided and as with the phono cartridge considerations covered in Sec. 9-6, a preamplifier must be included to use the dynamic microphones. The less-costly players will have provisions for crystal or ceramic microphone input only. If the input is for a dynamic microphone, the crystal types will overload circuitry unless dual inputs (crystal or dynamic) are provided.

Another input jack is for recording radio or phonograph programs. The more expensive units have additional heads for monitoring purposes. While a fixed-bias characteristic is common for the lower-priced units, a switch for changing the bias is often included. For best results with chromium-dioxide tapes, a different bias is required than for the iron oxide types.

Reference should be made to Sec. 1-5 and 1-6 for tape maintenance factors and phono equipment principles. If one channel's sound output is inferior to the others (or absent) control settings and switch setting should be checked. Also, the interconnecting coaxial cables linking the deck to the radio or amplifier should be tested for continuity or shorts. Also make sure the heads are clean because clogged heads are a common cause for poor performance. As discussed in Sec. 1-5 some demagnetizing may also be required.

Head wear can contribute to poor results, though if one channel gives much poorer performance than the other some circuit faults may exist. If tests point to a defective record or play head in a stereo machine (where one channel is affected) the leads to the two stereo head coils can be unsoldered and the wiring reversed. Upon playing, the head is at fault if the other channel now has the distortion or poor performance. If the same channel is defective the troubles must lie beyond the head itself. Check the switching circuits for open leads, and test circuitry for defective transistors and other components. Finally, check both the channels of the audio amplifier system used with the tape deck to see whether or not a defective stage in the external unit is contributing to the problem.

If the level meter is active during playback, but no sound is produced by that particular channel, the trouble must be in the resistive take-off network (R_5 and R_6), the feed line to the output jack or the jack itself (assuming the interconnecting cables from the deck to the amplifier

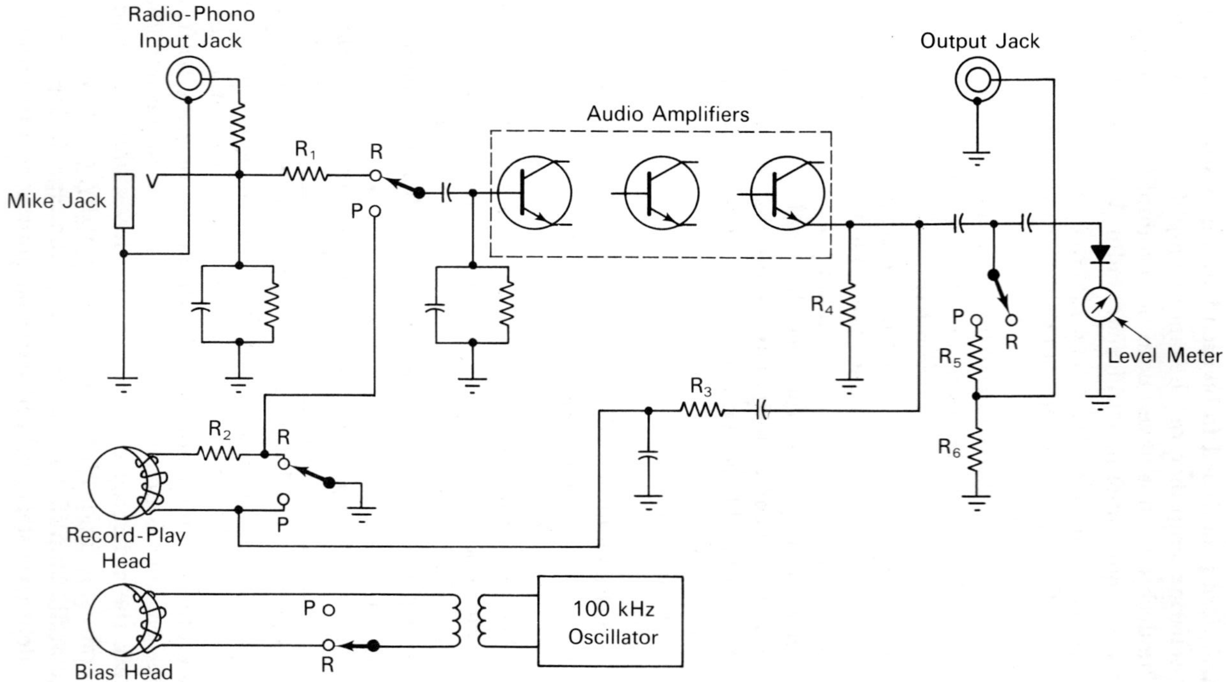

Figure 9-7 Tape deck switching system.

have checked out). Additional testing procedures for stereo amplifiers are given later in this chapter.

9-8 STEREO AMPLIFIER TROUBLESHOOTING

Unless both left and right channels of stereo become defective simultaneously, troubleshooting is easier for stereo amplifiers than for mono. If one channel has low volume or distortion or becomes inoperative, tests can be made on that channel and immediately compared with identical tests at the same points on the other channel. Obviously the same results should be obtained in simultaneous testing if the controls are set for a balance and if both channels are operating properly.

In many schematics the components having dual functions in the two channels are identified by special numbering. A typical example of this is shown in Fig. 9-8 where the left channel circuit shown at (Fig. 9-8A) has components numbered at a 200 count below the same parts for the right channel schematic (Fig. 9-8B). Thus, the volume control at A is designated as R_1 while the volume control section for B is numbered R_{201}. Similarly, the Q_1 emitter capacitor for the left channel is designated as C_3 while that for the right channel is C_{203}. Such numbering simplifies comparison checks because of easier identification of the equivalent part in each channel.

When voltages are given in the schematic as shown in Fig. 9-8, the bias factors and test methods discussed earlier are applied (see Sec. 3-8 and also Chapter 8). Note that Q_1 is an *npn* transistor, hence the forward bias between emitter and base should provide for a positive potential at the base with respect to the emitter (or, conversely, a negative emitter with respect to the base). The voltages shown, however, may be misleading if voltage-polarity relationships are not kept in mind. It would seem that the base is minus, since it has an indicated negative potential of 8 V. However, since the emitter is – 10 V compared to the base – 8 V, the base is *positive* with respect to the emitter by 2 V, satisfying the forward bias condition.

Similarly, for an *npn* transistor the collector should be positive with respect to the emitter. This condition is satisfied for Q_1 because the – 6 V at the collector makes this terminal positive by 4 V with respect to the – 10 V emitter. These relationships also prevail for Q_2 where the – 12 V at the base makes this point positive with respect to the – 15 V emitter. At the collector the – 10 V makes it positive with respect to the – 15 V at the emitter.

In many circuit boards troubleshooting and testing procedures are simplified because the component parts are identified (sometimes on

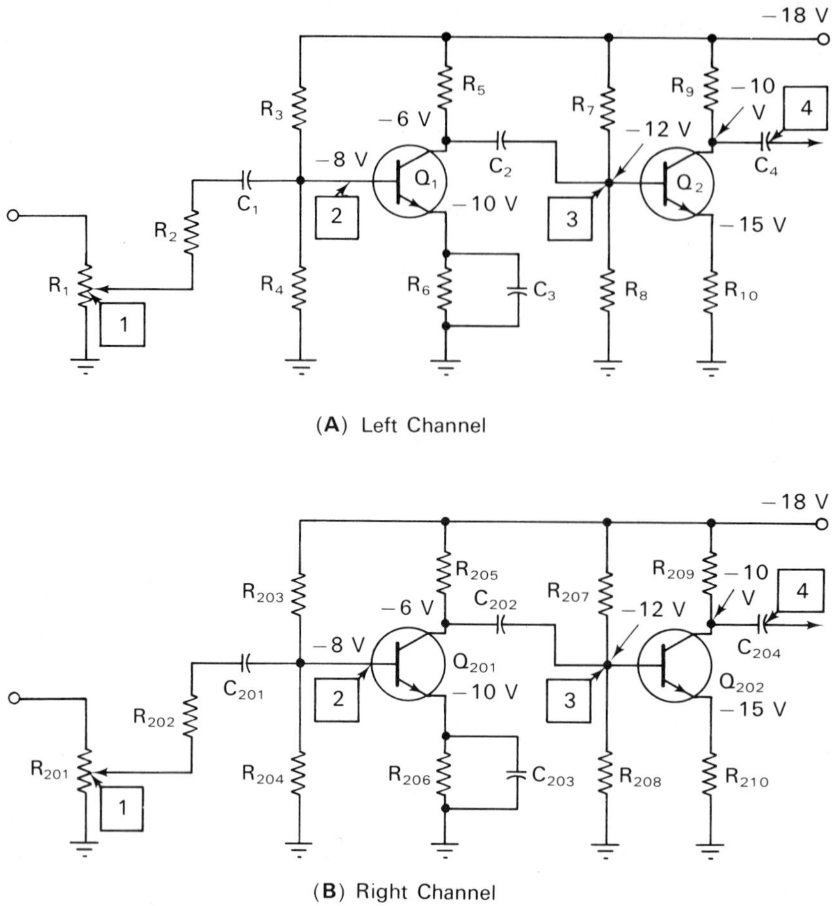

Figure 9-8 Stereo channel comparisons.

one side of the board only, and on occasion both sides). The circuit board sections illustrated in Fig. 9-9 indicate the methods usually used to aid in identification. The underside of the printed-circuit board is shown in Fig. 9-9A. Symbols are printed on the underside to correspond to the placement of the actual component on the topside of the board. Thus, by comparing these symbols with the schematic (Fig. 9-8 in this instance) parts location is simplified. Often the topside where the actual components are wired in have identifying letters and numbers printed as shown in Fig. 9-9B. Thus, parts locations for both top and bottom for both stereo channels are aided considerably.

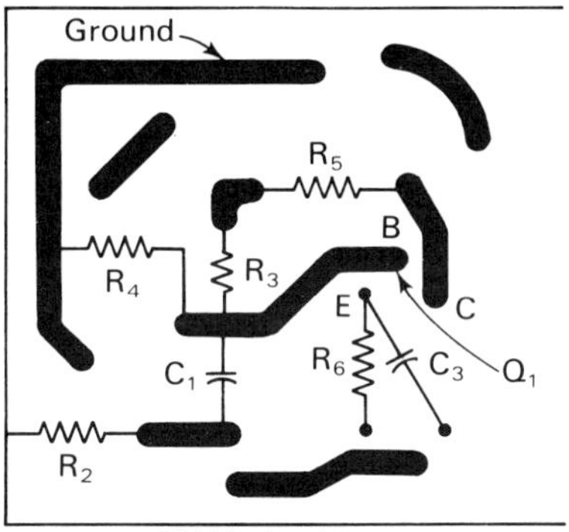

(**A**) Underside

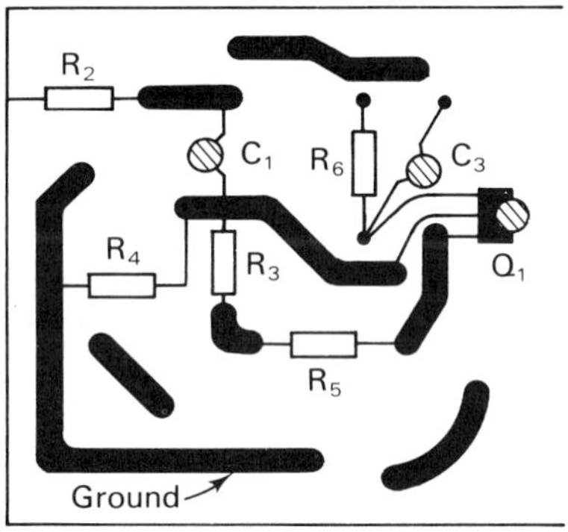

(**B**) Topside

Figure 9-9 Parts identification on circuit board.

After specific components or test points have been identified on the chassis, voltage checks and comparisons can be made in succession from the input circuitry through to the output. As soon as a difference in voltage is noticed between the defective channel and the one operating properly the power is shut off and components are tested for shorted or open resistors, resistors that have changed in value, defective capacitors or faulty transistors. Some minor voltage differences will occur, however, because of slight variations in transistor characteristics or parts. Generally the tolerance range for resistors and capacitors will be specified in the schematic and often a ten percent variation in resistance is permitted. If some resistors require a closer tolerance (five percent or less) they are usually so identified and care must be taken to replace the unit with one within the tolerance range.

As discussed in Chapter 4, an oscilloscope can be used for testing the relative gain of amplifier stages and to ascertain if the amplifier is introducing signal distortion. A signal from a generator is applied to the input of the amplifier, and a scope placed at the successive test points (numbered squares) shown in Fig. 9-8. The patterns observed are then compared to those obtained for the second channel. If a dual-trace scope is available, comparisons between the two channels can be made without having to switch scope input leads between the two. A typical dual-trace scope display is shown in Fig. 9-10 where the same amplitude square wave into each channel shows a decline in output for the right channel.

Because the balance control was at center and the same amplitude signal introduced into each channel, the results shown in Fig. 9-10 call for a check of voltages and components for the right channel. The dual-trace pattern can be picked up at the successive test points to find out where the trouble occurs. Thus, if the amplitude difference in the square waves does not appear at test point #3 but does show at test point #4, transistor Q_{202} should be checked in addition to testing for correct resistance values of R_{207}, R_{208}, R_{209}, and R_{210}. Coupling capacitor C_{204} should also be tested for leakage or short-circuit conditions.

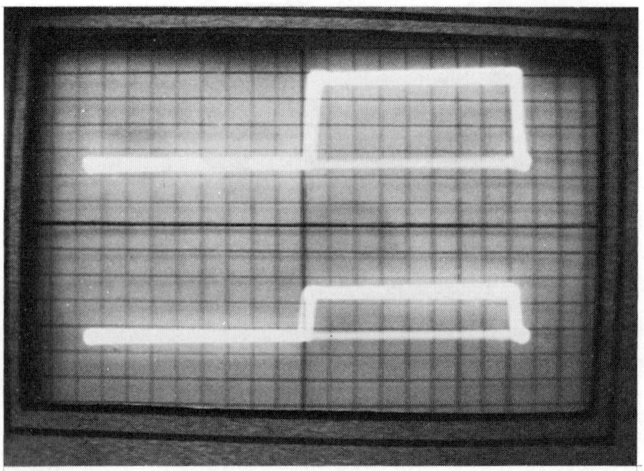

Figure 9-10 Square-wave test of stereo.

Note that the square waves of Fig. 9-10 do not exhibit any distortion, indicating that the only fault in the right channel appears to be a gain reduction. For comparison, note the square-wave pattern of Fig. 9-11, which was injected into an amplifier having a faulty transistor that diminishes high-frequency components. The distortion introduced into the waveform is shown in Fig. 9-12 where the trailing edge of the square wave has lost its sharpness (high-frequency component). (Also compare Fig. 4-11 and 4-13 to Fig. 9-11 and 9-12.)

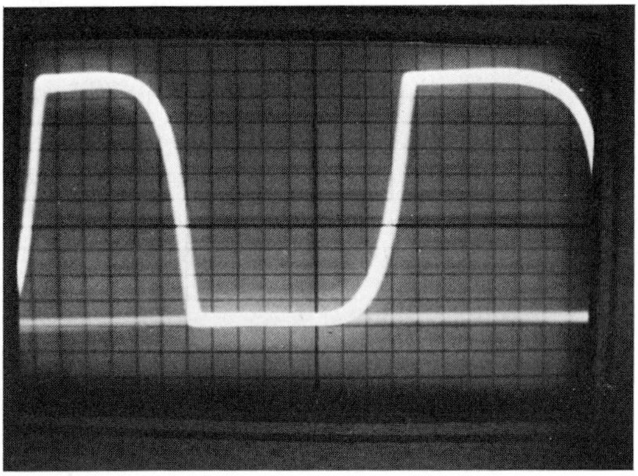

Figure 9-11 Square wave input.

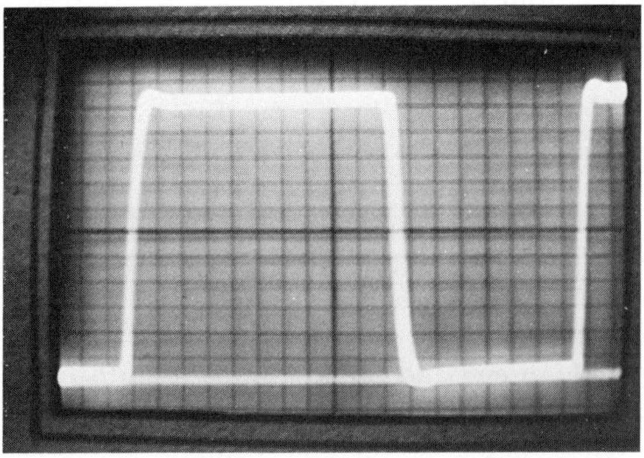

Figure 9-12 Output distortion.

9-9 HI-FI RELATED TERMS AND DESIGNATIONS

A number of terms and designations are used in stereo and quadraphonic descriptions to indicate the merits of the equipment relative to certain standards. While these are more useful when purchasing new equipment (or when designing systems during manufacture) than they are in testing and troubleshooting, the technician should be familiar with them so that faulty equipment can be restored to the peak performance it had originally (see also Section 3-6 and 3-7). Specialized test equipment is available (sensitivity meters, distortion meters) for laboratory-type testing of amplifier characteristics, though such special test features are often incorporated in the more expensive models of the test equipment described in earlier chapters.

Universal agreement among equipment manufacturers and government agencies has not always been achieved easily and differences in opinions still exist. General standards have been attained in a number of areas, however, though it must be kept in mind that a listing of performance characteristics by a manufacturer tends to favor that manufacturer. Thus, after servicing stereo and quadraphonic equipment or hi-fi amplifiers, the specification claims of the manufacturer may not always be achieved at the levels listed in equipment advertisement. Thus, if the equipment appears to function normally and gives good performance after repairs, the technician need not be too concerned if tests do not indicate the peak performance specified by the manufacturer. If some doubt exists, tests should be performed on identical (but new) equipment to ascertain the actual performance characteristics.

Total Harmonic Distortion (THD). This is a comparison of the waveshape of the output signal versus that of the input, usually based on a 1000 μV input signal to a receiver. Distortion in modern receivers is kept well below the point where it is noticeable even to an experienced listener. Some manufacturers may give the THD for several frequencies, or the total frequency span for the receiver (such as 0.3 percent for 20 to 20 kHz) (see Sec. 4-3 and Figs. 4-9 through 4-13).

In distortion rating comparisons, the lower figure denotes the preferred rating. Thus, a rating of 0.32 percent at 1000 Hz is better than 0.5 percent at 1000 kHz. THD based on full-frequency response is preferred, since distortion tendencies are greater at the extremes of the frequency range, both at the high end and the low end. Also, distortion tends to be somewhat greater for stereo than mono (and the signal-to-noise ratio lower for a given radiated power and without special noise-reduction systems in operation).

IHF Sensitivity. The IHF portion of the term refers to the Institute of High Fidelity and the sensitivity rating relates to the tuner characteristics of the radio. It indicates the μV of signal required at the input terminals of the receiver (antenna input) to produce an audio output of 30 dB greater than the general background noise. Some manufacturers use a 50 dB reference as being more realistic. The FM tuner IHF sensitivity is better for the lower figure. Thus, a 1.8 μV sensitivity is better than a 1.9 μV, and much superior to a 2.3 μV rating.

Signal-to-noise Ratio (S/N). This relates to the degree by which a receiver raises the signal amplitude above the general residual noise level. The higher the S/N ratio the better. Thus, a 70 dB rating is superior to a rating of 65 dB.

Capture Ratio. This term relates to the tuner's ability to pull in a station (*capture* it) over other stations to which it is not tuned. Thus, the capture ratio defines that characteristic of the FM tuner which rejects nearby stations, or more powerful distant ones, while passing through only the station to which it is tuned. Capture ratio is designated in dB and the lower ratio is preferred. Thus, a capture ratio of 1 dB indicates a rejection system superior to that of a receiver rated as having a capture ratio of 2 or 3 dB (see also the discussion on selectivity which follows).

Selectivity. Selectivity relates to the steepness of the bandpass characteristics of the tuner and IF stages (see Figs. 7-1, 7-2 and 7-15 plus related discussions). For selectivity ratings of a receiver the higher the number the better the bandpass characteristics. Thus, a rating of 80 dB is superior to one of 60 or 40 dB.

Stereo Separation. This specification relates to the ability of a receiver, stereo amplifier or stereo phono cartridge to minimize the leakage of signals from one channel to the other. Generally, a 20 dB rating provides for adequate separation and specs are usually based at a test made for a 1000 Hz signal. The higher the rating, the better the separation. Thus, a 25 dB rating indicates a superior characteristic over a unit having a 20 or 15 dB rating.

Power Output. The power output in watts relates to the voltage-times-current characteristics available to drive loudspeakers (which consume the electric energy generated). Power output depends on the amplitude of the signal input and the ability of the circuits to raise such signal levels to that required.

Power output ratings for amplifiers have been given in *rms* values, EIA and IHF. The *rms* relates to *root-mean-square* values, also known as the *effective* value of voltage, current or power, which is obtained by multiplying the peak value of an ac signal by 0.707. It is that power that is equivalent to dc value needed to do the same work as the ac value. (Thus, if the peak ac amplitude coming into a home is 165 V, the rms or effective value is $0.707 \times 165 = 116.655$ V.) In laboratory testing and rating of this type the audio power is measured by using a single-frequency signal input and measuring the output rms voltage across a given value of load resistance ($P = E^2/R$).

EIA refers to Electronic Industries Association and IHF to Institute of High Fidelity, and the power rating standards of each are basically similar and based on testing a preconditioned amplifier, that is, one that has been turned on for a specified period of time to reach normal operating temperatures. Power ratings usually refer to the wattage available per channel. Thus, a rating of 40 W per channel provides a total output power of 80 watts (for a given sinewave signal input wherein the voltage is constant and the volume controls are set for maximum gain). Higher ratings provide for greater power for additional speaker systems.

Anti-skate Device. An anti-skate device is a unit incorporated into some record players to balance out the lateral force applied to a phonograph needle (and its tone arm) as the grooves move the assembly toward the record center during play. It minimizes *skating* (sliding of needle) over the record surface (see Sec. 1-6).

Tracking Force. Tracking force is the grams of pressure exerted by a phonograph needle on the record during play (see Sec. 1-6). A lower tracking force produces less record wear.

Compliance. This is another term relating to phonograph needle-cartridge operation. It refers to the ability of the phonograph needle to *comply* with the variations of the record groves (see Sec. 1-6). The greater the compliance, the better the reproduction performance of various frequency signals.

9-10 BASICS OF QUADRAPHONIC SOUND

In contrast to stereo where two amplifiers and two speakers are needed, the quadraphonic system is a four-channel one requiring four individual amplifiers (two stereo amplifiers) and four separate speakers. Thus, there are two front speakers and two rear, with full capabilities for surrounding the listener with individual sound sources. Any two speakers will provide for stereo effect, that is, an instrument can appear to come from speaker 1, from speaker 2, or any point between 1 and 2 if proper phasing exists. Thus, the phasing factors covered in Sec. 9-6 must be followed in connecting the speakers to quadraphonic systems.

Several systems have been developed, each differing somewhat from the other in the manner of signal mixing, phasing relationships and decoding principles. One method for phono discs is known as CD-4 and requires a demodulator system for proper decoding of the phono record music. Another system is known as SQ, and requires a matrix and decoder type system for proper playback. The matrix sections consist of resistor and capacitor combinations for selecting certain signal amplitudes and phasings. An encoder may use an IC module for the mixing and signal-processing unit.

With the demodulator or decoder systems true quadraphonic (four-channel) reproduction is obtained. There are some systems available, however, that provide for inexpensive four-channel effects without actually forming the quadraphonic system. Such systems are known as *synthesized* four-channel sound. In such a method the stereo signals are processed by resistor-mixing circuitry to give the illusion of quadraphonic reproduction by introducing a slight difference in sound reproduction between front and rear speakers.

Four-channel FM radios are commercially available with built-in decoders for both the broadcast quadraphonic transmissions and the tape and phono disc recordings. Standard stereo (or mono), of course, can also be received with these sets. Some of the lower-priced stereo receivers have provisions for adding decoders to them. They may have provisions for synthesized four-channel sound and contain four separate amplifier sections.

Separate four-channel decoders are also available for use with an existing stereo receiver or amplifier, as shown in Fig. 9-13. For quadraphonic operation, however, an additional stereo amplifier must be used. The four speakers, two stereo amplifiers, a decoder unit, plus phono and tape decks make for a complex intercable connection array. The phono connections are not shown for Fig. 9-13, though these would increase the number of interconnecting cables, of course, as would an additional tape recorder such as the reel-to-reel type or the cassette.

The auxiliary output of the receiver (sometimes marked *tape output*) is fed into the decoder-matrix unit as shown, while the decoder also feeds signals to the radio for front channel reception of tape. Another output from the decoder supplies the signals to the second stereo amplifier for the two rear speakers. The quadraphonic tape player outputs are fed to special inputs on the decoder unit as shown.

Because of the many interconnections, any loss of sound from one section calls for a check of the cables to ascertain that they are tightly plugged in and that no open circuit exists. Testing procedures given earlier in this chapter for stereo also apply to the assembly illustrated in Fig. 9-13.

Absolute separation between each channel is not always achieved with some recorded tapes or discs and this factor must be kept in mind when testing such assemblies. Also, make certain that balance and volume controls for rear and front speakers are properly adjusted. Phasing for the four speakers must be correct as mentioned earlier. Test records are available for proper balancing, phasing and general-set up procedures.

The decoder make-up consists of switching circuits using push-button or rotating knob controls. As shown in Fig. 9-14, the switches interconnect the input and output lines with the resistor-capacitor matrix

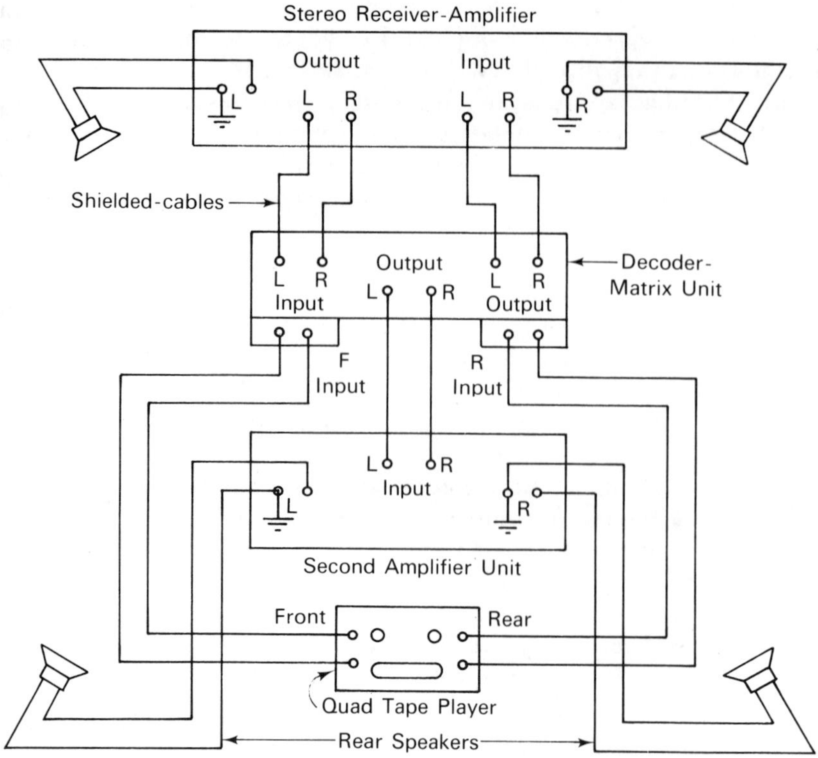

Figure 9-13 Basic Quadraphonic system.

circuitry. Some commercial types use the integrated-circuit module containing the necessary signal-mixing and solid-state processing units.

Very little trouble is experienced with the matrix circuitry because of low voltages used. Most of the servicing problems occur in the switches (either slide or rotary) and these should be checked initially when problems arise. Because of circuit variations among manufacturers, the service notes and schematic for a particular model should be consulted if component replacement is required. Resistors and capacitors in the matrix unit are color-coded and have standard values (hence are easily replaced), but switches may be special types. For proper fit and correct output terminals, therefore, it may be necessary to obtain factory replacement parts. Also, the IC module failure requires a direct factory replacement because standard replacements may be difficult to find.

As mentioned earlier in this chapter, failure of a single channel may also be caused by defects in the particular amplifier associated with the channel, as well as the speaker and speaker wiring. Also, troubles could occur in the stereo receiver/amplifier where one channel could fail

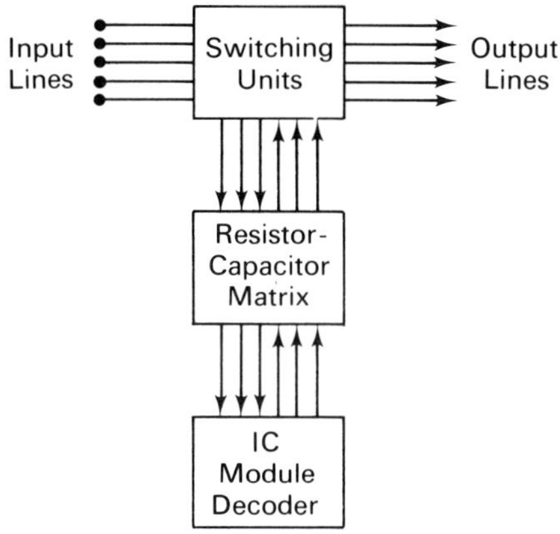

Figure 9-14 Decoder interconnections.

and thus affect quadraphonic operation. Similarly, a defective channel may be caused by troubles in the phonograph or tape player units attached to the quadraphonic assembly, so these must also be checked for possible open or short circuits or component failures.

General factors regarding phono cartridges, tape heads, and other accessories discussed in Chapter 1 and other sections herein, also apply to quadraphonic. Each audio channel supplied by the matrix-decoder system of the receiver (or the separate unit) is an individual basic mono system and can be treated as such when it behaves in an intermittent fashion, or has low output or distortion. After it has been repaired, a mono recording can be used to check the output from each channel for equal-level output. (A scope provides a better and more accurate check than just listening to the quadraphonic sound.) In tests of this type, however, again ascertain that the balance controls and volume controls are properly set among the channels.

10

television receiver sections and adjustments

10-1 TV TEST PRECAUTIONS

Basic tests and measurements on television circuitry are similar to those performed for other electronic circuitry and consist of voltage reading, resistance and continuity checking and, on occasion, current readings. Because of the number of circuits in a television receiver, however, testing procedures are often more involved and critical. This is normally true for color receivers where additional circuits are involved. Also, in television, a greater variety of signals must be handled compared to radio receivers or high-fidelity equipment. Such signals include blanking pulses, sync pulses (both at a horizontal and a vertical sweep rate), video signals (amplitude modulated), sound signals (frequency modulated), sideband components of color, plus a burst signal for color-carrier oscillator synchronization. As shown in Fig. 7-4 and Fig. 7-5, a wide band is necessary to convey both the picture and sound signals, hence special test equipment must be used to measure and align the various stages.

In addition to the low voltages found in radio and high-fidelity equipment, high voltages are also present in television (as high as 30 kV in color sets). Consequently, precautions must be taken in testing and measuring the various circuits making up the horizontal output, the focus rectifier, the damper and the high-voltage circuits. Special probes must be used as detailed later in this chapter. Reference should also be made to earlier discussions, particularly as they relate to television circuit testing (section 1-1, 1-4, 3-8, 3-10, 4-6, 6-5 and others).

The high-voltage system in television usually employs high-value series resistances so that a rise in current will create a large drop across the resistor. This is a safety feature that minimizes the danger of heavy current flow through the body upon contact with the high voltage. A severe shock can be encountered, however, and could prove fatal to one whose body resistance is low and who has had (or has at present) heart trouble. Thus, extreme caution should be exercised. Additional precautions are given later in this chapter.

Another hazard that may be present whenever voltages are in excess of 15 kV is the danger to x-ray exposure. Modern receivers generally have shielding adequate to decrease the dangers of exposure, though the precautions indicated by the manufacturer should be strictly observed. In particular, when working with older receivers, try to make checks and tests with the power shut off when working on the chassis. When cages and shields are removed from high-voltage or horizontal-output circuits to facilitate testing, avoid prolonged exposures when the set is turned on.

10-2 BLOCK DIAGRAM OF TV

A block diagram of the basic circuits present in a color television receiver is shown in Fig. 10-1. Those circuits that are exclusive to color receivers are indicated by a double line at the top of the block. Thus, for a black and white receiver, only those blocks would be used that have no double bar at the top. As this block diagram shows, the color receiver uses all the circuits contained in a black and white receiver, plus the special circuits needed to process the color signals.

Many of the circuits shown in the blocks have already been discussed and illustrated in earlier chapters with reference to tests, measurements, continuity testing and alignment. Additional television circuitry is covered in this chapter. Of course, tube circuits can be formed from the transistor types illustrated, with a basic circuit function and purpose the same.

A brief explanation of the type signals comprising television transmission will help in understanding the purpose for the circuitry shown in Fig. 10-1. In amplitude modulation as used in radio, the carrier signal (with sidebands) varies in amplitude as shown in Fig. 10-2A. If the carrier

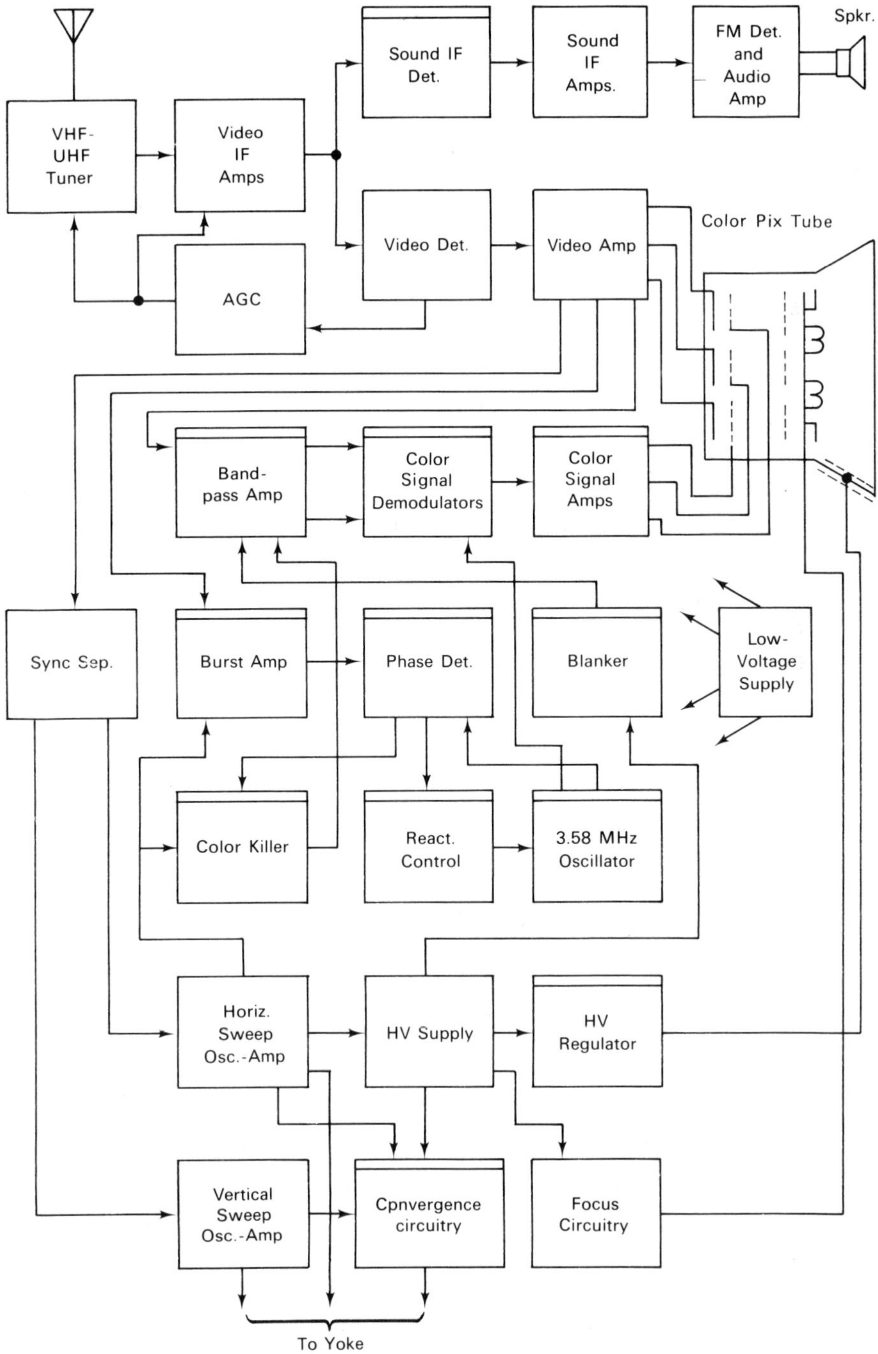

Figure 10-1 TV receiver circuitry.

is 900 kHz and the modulating tone 400 Hz, the carrier amplitude would vary 400 times per second. Various modulating signals would change the amplitude variations accordingly.

In television, the picture signal is also amplitude modulated, so that the carrier (plus sidebands) appears as shown in Fig. 10-2B. Here, however, certain definite amplitude peaks are reached periodically, in contrast to the random amplitude changes shown in Fig. 10-2A. The re-

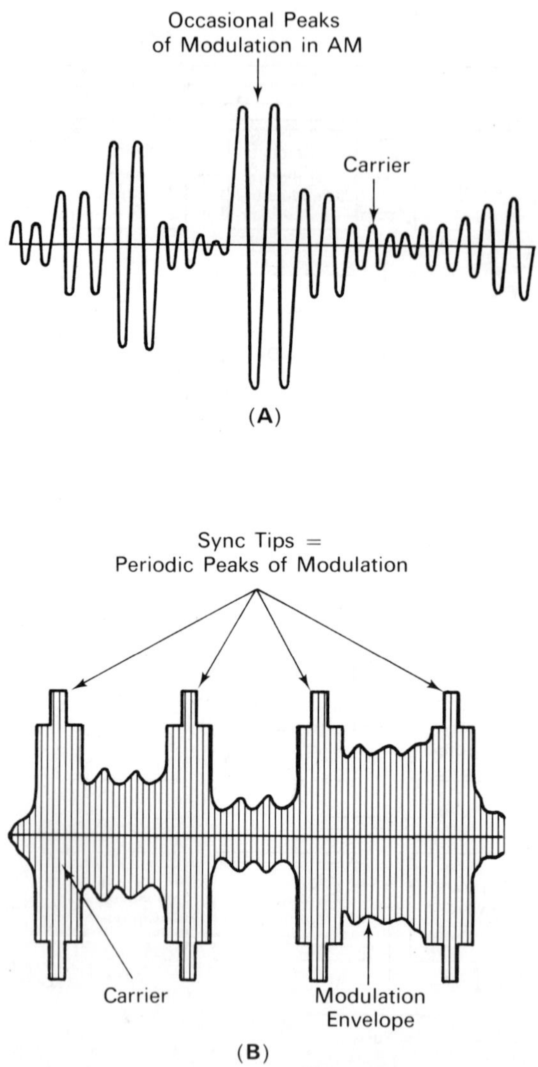

Figure 10-2 AM radio vs. television signals.

curring peaks are produced by the synchronization and blanking pulses that are transmitted at two rates, 15,750 pps and 60 pps. The 15,750 is used for the horizontal beam trace across the tube, and the 60 rate for the vertical downward sweep of the beam. Thus, the horizontal sweep rate of a scope can be set (either at 15,750 or 60) to lock in these pulses during testing (see Figures 6-8 and 6-9).

When the video signals are demodulated at the detector the resultant signals appear as shown in Fig. 10-3. Figure 10-3A illustrates the horizontal blanking pulses with the horizontal sync pulses mounted on them. Between the blanking pulses are the video signals, which vary the brightness of the electron beam in the picture tube to trace out the picture on the screen. Figure 10-3B shows the relative phase of the signal

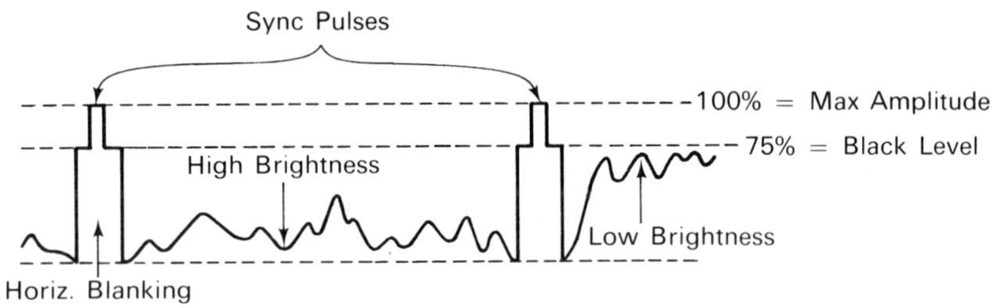

(A) Demodulated Video Signal

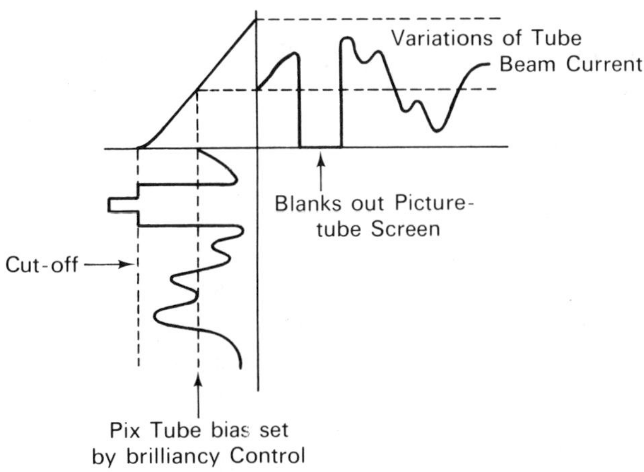

(B) Video Signal Applied to Picture Tube

Figure 10-3 TV video signal characteristics.

at the picture tube, which blanks out the screen or produces different brightness levels.

In color transmission the carrier is suppressed and only the sidebands are received. Thus, the receiver must generate the carrier so it can be combined with the color sidebands and detected. To lock in the picture color carrier generated by the receiver (3.58 MHz) a special burst of about 8 cycles of signal is sent along with the blanking pulse, as shown in Fig. 10-4. This burst signal is used as detailed later in this section.

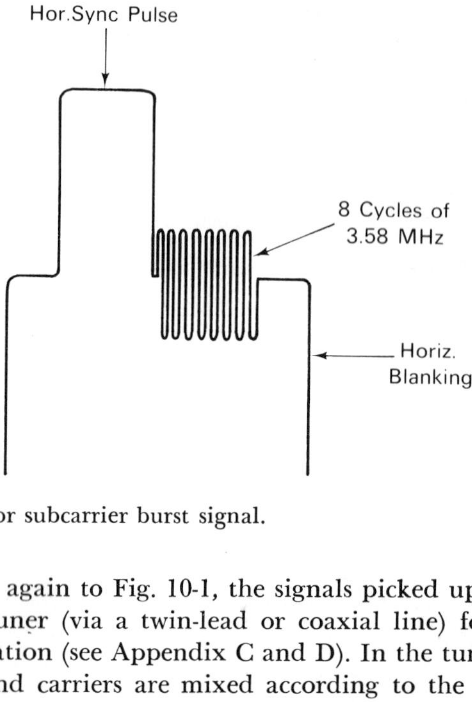

Figure 10-4 Color subcarrier burst signal.

Referring again to Fig. 10-1, the signals picked up by the antenna are fed to the tuner (via a twin-lead or coaxial line) for selection of a VHF or UHF station (see Appendix C and D). In the tuner the incoming picture and sound carriers are mixed according to the superheterodyne principle discussed in Sec. 6-3. Two intermediate frequencies are produced, one having a frequency of 45.75 MHz (video) and the other (the sound IF) having a frequency of 41.25 MHz (as was shown in Fig. 7-5). Several video IF amplifiers increase the gain of these signals and apply them to the detector system. The video detector derives the video information, which is applied to a video amplifier stage and to the picture tube.

In television receivers another mixing process occurs in the detector system, where the 45.75 MHz signal heterodynes with the 41.25 MHz signal to produce another sound IF of 4.5 MHz. This sound IF is the same for either black-and-white receivers or color sets. With black and white receivers, the 4.5 MHz signal is obtained in the video detector and sent to the sound IF amplifiers. In color sets, however, a separate detector obtains the sound IF because the bandpass utilized would not provide enough gain. After sound IF amplification, a standard FM detector obtains the sound signals and after a stage or two of amplification, the audio is applied to the loudspeaker.

Typical video detector circuitry was shown in Fig. 6-6. A video amplifier circuit is illustrated in Fig. 5-6, and FM detector circuitry is shown in Fig. 6-5. Response curves and traps were shown in Figures 7-4 through 7-7. In color receivers, the output from the video amplifiers that feeds the cathodes of the color tube consists of the black-and-white signal, also termed *luminance or Y* signal (see Appendix E).

The color sideband signals from the video amplifier system are fed to a bandpass amplifier, which rejects all signals except the desired color sidebands. A 3.58 MHz oscillator generates a color subcarrier signal that is compared to the 3.58 MHz burst that comes in on the horizontal blanking pulse, as shown in Fig. 10-4. A phase detector system is used to make comparisons and a reactance control circuit regulates the frequency of the oscillator if it should drift away from the incoming sync burst (see also Fig. 4-19 and 4-22). The burst amplifier is triggered by a pulse from the horizontal sweep circuit so it passes a signal at the precise time that the 3.58 MHz burst appears on the blanking pulse. The color killer prevents color signals from appearing on the screen during black and white picture reception. The killer function occurs when no 3.58 MHz sync burst appears at the burst amplifier.

Color signal demodulators detect the color signals from the carrier and sidebands (see Fig. 3-11) and apply these signals to respective blue, green and red amplifiers. The output is applied to the grid circuits of the color tube as shown in Fig. 10-1. Thus, as shown later in this chapter, the three color signals produce amplitude variations of the electron beams within the tube and cause respective phosphorus dots on the tube face to glow with an intensity depending on beam velocity. All three colors are necessary to produce white areas in the televised scene, plus the various shades of color. If any particular color amplifier does not function it would be impossible to obtain a black-and-white picture, or true colors (except the primary colors of the two amplifiers that still function).

As also shown in Fig. 10-1, a sync separator obtains the individual horizontal and vertical sync pulses from the video signal and applies them to the vertical and horizontal sweep circuits. These sweep circuits generate the sawtooth signals that sweep the electron tube beam across and down the face of the picture tube. The high-voltage supply is part of the horizontal output system (see Fig. 2-7A) In color receivers a high-voltage regulator circuit is used to stabilize the output, and a convergence circuitry is used for three-gun tubes to assure that proper registration is obtained for the three beams, so each falls exactly on the proper matching color dot. Additional details on these processes are covered later in this chapter.

In color receivers the horizontal sweep rate is 15,734 MHz instead of the 15,750 for black and white. Also, for color, the vertical sweep rate is 59.94 Hz instead of 60 Hz. In both instances, however, the two vertical and two horizontal frequencies are sufficiently close to assure good lock-in when receiving either color or black-and-white transmission. The frequency differences are necessary to accommodate the color signals within the previously black-and-white transmission bandwidth without causing interference.

The horizontal sweep of the beam within a picture tube traces a white line (in the absence of a picture) and the vertical sweep pulls the horizontal sweep downward. The white glow thus formed on the screen is termed a *raster* and this is transmitted in a 3 by 4 ratio. Thus, if a picture is 16 inches wide, it would be 12 inches across. A dark screen (no raster) would indicate loss of high voltage, brilliancy control turned too far down or defects in the brilliancy circuitry. If sound is absent also, there may be open circuits in the ac line, a blown fuse or a defective on-off switch.

10-3 COLOR TV DETECTOR AND VIDEO AMPLIFIER

A typical sound-video detector system for color television is shown in Fig. 10-5A. The upper diode detector does *not* produce the sound signals (audio), but rather the sound intermediate-frequency signal of 4.5 MHz. This IF signal must be demodulated again in a standard FM detector to produce the audio signals (see Fig. 6-5).

The video detector circuit obtains the picture signals as in Fig. 10-3. On an oscilloscope the video signals appear as shown in Fig. 6-8. To keep the 41.25 MHz sound IF from appearing on the picture-tube screen a series-resonant circuit trap is employed (C_6 and L_2). Adjustment of the tuned transformer core eliminates the interference from sound lines on the screen by providing a low impedance shunt path to ground at resonance.

Failure of the sound detector diode would prevent audio output, but a picture would still appear on the screen. Failure of the video detector diode would result in a raster only, with no picture appearing. Sound output, however, would still be present. It is unlikely that both diodes would become defective at once, but this is possible. In such an instance there would be no sound or picture, but a raster would still be

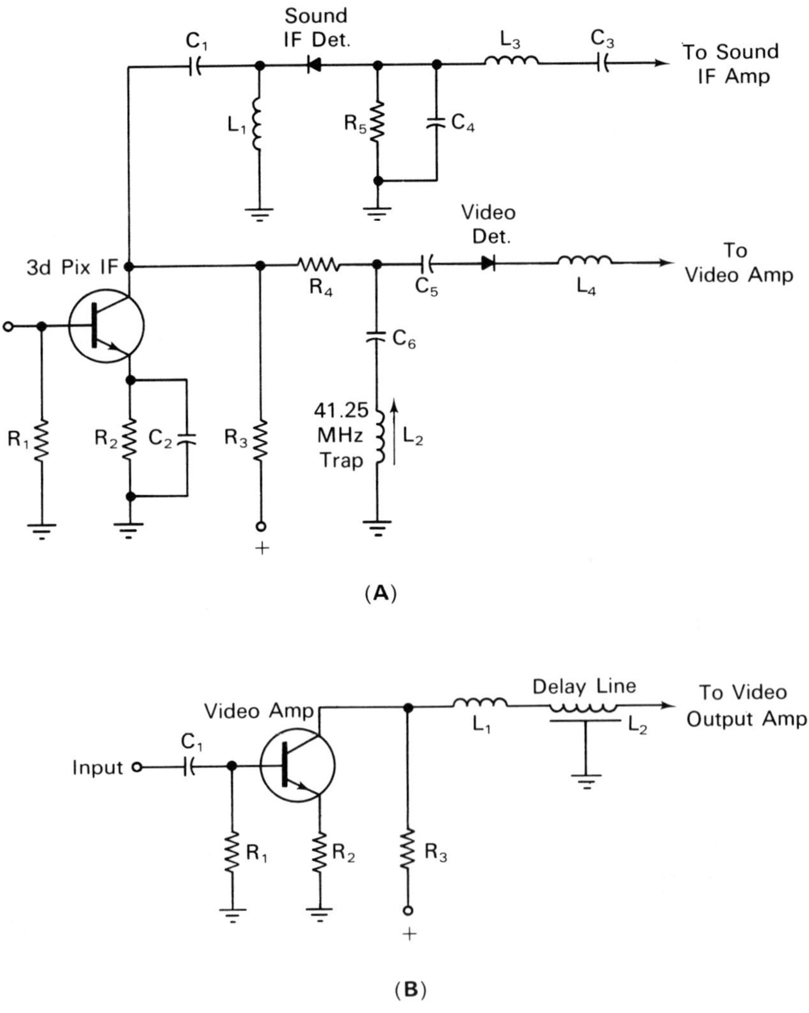

Figure 10-5 a. Sound and video detectors.

b. Basic video amplifier in color TV.

present. As mentioned in Sec. 10-2, the absence of a raster (a dark screen) is caused by other troubles.

A typical video amplifier is shown in Fig. 10-5B. Inductor L_1 is a peaking coil and its function is described in Sec. 5-4. The delay line is necessary to compensate for the delay encountered in the color circuits because of bandpass characteristics. Because the color signals are delayed slightly, the luminance (Y) is delayed in the same amount so as to maintain color detail synchronization with the black-and-white signal components. (See also Figures 5-6 and 6-6 for black-and-white circuit versions of those shown in Fig. 10-5.)

10-4 COLOR KILLER CIRCUITRY

As mentioned in Sec. 10-2, the color killer circuit prevents signal entry into the color circuits during reception of a black and white picture. This process eliminates the interference that would result if video signals pass through the color circuits during this time. A typical killer circuit is shown in Fig. 10-6.

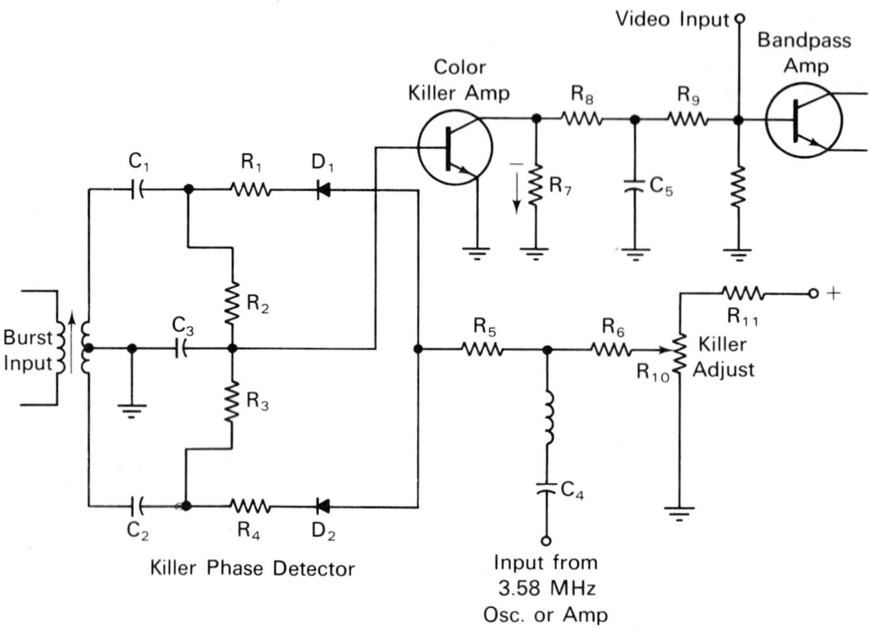

Figure 10-6 Color killer circuitry.

The killer phase detector applies zero (or reverse) bias to the base of the color killer amplifier transistor, preventing conduction during the presence of a burst-signal input to the phase detector. The bias can be set by adjustment of the R_{10} variable resistor, also known as the *killer threshold control.*

During black and white reception, no 3.58 MHz burst arrives at the input to the phase detector, and an unbalance occurs. The detector now applies a positive voltage to the base of the killer (forward bias) and causes conduction. The rise in voltage across R_2 produces an increasingly negative potential at the base of the bandpass amplifier. The negative base is thus a form of reverse bias, and the bandpass amplifier goes into nonconduction, preventing signal entry from the video input line.

For receivers having a variable killer threshold control, adjustments are made with the receiver on, but not tuned to a station (or with the antenna disconnected). The color and tint controls should be set at midrange and the killer control at the minimum resistance point. Colored noise interference (snow) should appear on the screen.

The killer control is now turned up slowly until a point is reached at which the colored interference no longer appears (or is at a minimum level). The adjustment should be just enough to remove the snow interference. Check the result by tuning to a station transmitting color and rotate the tint control through its full range. If the tint control is capable of removing all color, the threshold control setting may be too high and readjustment may be required. Consult the service notes, if available, for deviations from the foregoing procedures if troubles still persist.

General checks on the circuitry are required if poor performance (snow interference) is present. Initially, however, make certain that the antenna and transmission line system is supplying a good signal input to the receiver. In a weak signal area (or if open lines and antennas are present) interference will be present, of course, regardless of the killer setting.

For component troubles, check to see that a symmetrical phase detector circuit is present. Resistors R_2 and R_3 should match in ohmic value, as should resistors R_1 and R_4. Capacitor C_1 and C_2 should have the same value and present an open circuit for dc. Diodes D_1 and D_2 should be checked for proper unidirectional characteristics. Obviously, if one is shorted or open, the entire circuit function will be affected.

10-5 BANDPASS AMPLIFIER

As noted in Sec. 10-2 the bandpass amplifier has resonant tuned circuits selective to the desired color-sideband signals but rejecting those above and below the established bandpass. In addition the bandpass amplifier prevents passage of pulse and sync information into the color circuits (which would cause interference on the screen). Elimination of the sync pulses is accomplished by periodically gating the bandpass amplifier transistor (or tube, if used) into nonconduction.

A typical bandpass amplifier is shown in Fig. 10-7. The signal from the video amplifier is applied to the chroma take-off inductor L_1, which has an adjustable core for tuning purposes. A blanking-pulse input is obtained from the horizontal output transformer and applied to the junction of the two series tuning capacitors as shown. The blanking voltage opposes the polarity of the power supply voltage, hence the short-duration blanking pulse momentarily shuts off the transistor current flow in the emitter-collector circuits. Thus, the blanking pulse (synchronized by the incoming sync signal) shuts off the transistor precisely at the time sync pulses would be present at the input.

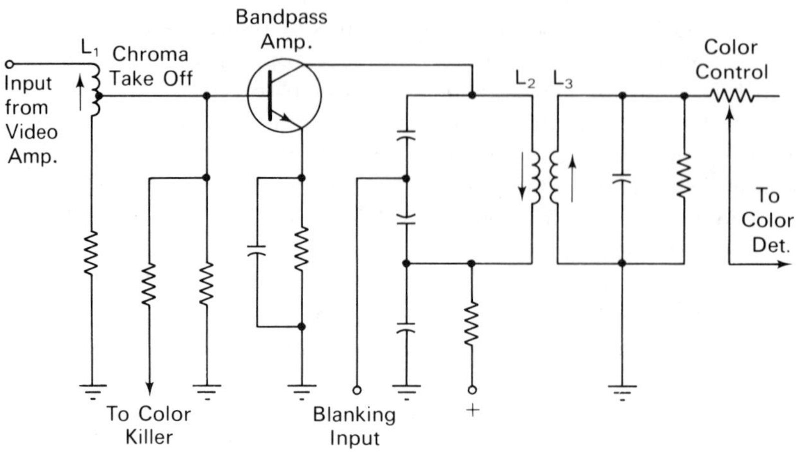

Figure 10-7 Bandpass amplifier.

The capacitors shunting L_2 and L_3 form resonant circuits around the 3.58 MHz region to accommodate the chrominance sideband signals (from about 3 MHz to 4 MHz). Improperly tuned input and output circuits will result in a decrease or loss of color. It must be emphasized, however, that lack of color does not necessarily mean improperly tuned circuitry. Before tuning is attempted, adjustments of the color control should be tried and general trouble-shooting procedures undertaken as detailed for other circuits earlier. Such tests should include continuity checking, component value-testing and voltage readings. Signal tracing can also be undertaken, plus vectorscope testing as discussed in Chapter 11.

Since the bandpass amplifier requires a blanking pulse input for proper function, an oscilloscope can be used to check for the presence of this pulse. Without the pulse, the sync information riding into the color circuits will cause streaking and variable-line interference on the screen. If alignment is required, use a sweep generator and marker, plus a diode (detector) probe for the oscilloscope (see Figures 7-12 and 7-16).

10-6 SYNC SEPARATOR

The sync separator circuit is common to both the black-and-white receivers and color sets. The separator, as the name implies, separates the sync (both horizontal and vertical) from the video information and passes the sync pulses on to the vertical oscillator and the horizontal oscillator to lock them into synchronization with the horizontal and vertical sweep frequencies accompanying the video information.

A typical sync separator stage is shown in Fig. 10-8. With an *npn* transistor a positive-polarity signal is applied to the base as shown. Since no forward bias is present for the transistor, no conduction occurs until the video signal arrives. The positive-polarity signal provides the necessary bias to cause conduction upon arrival of the high-amplitude sync. Hence, only the sync signals pass through.

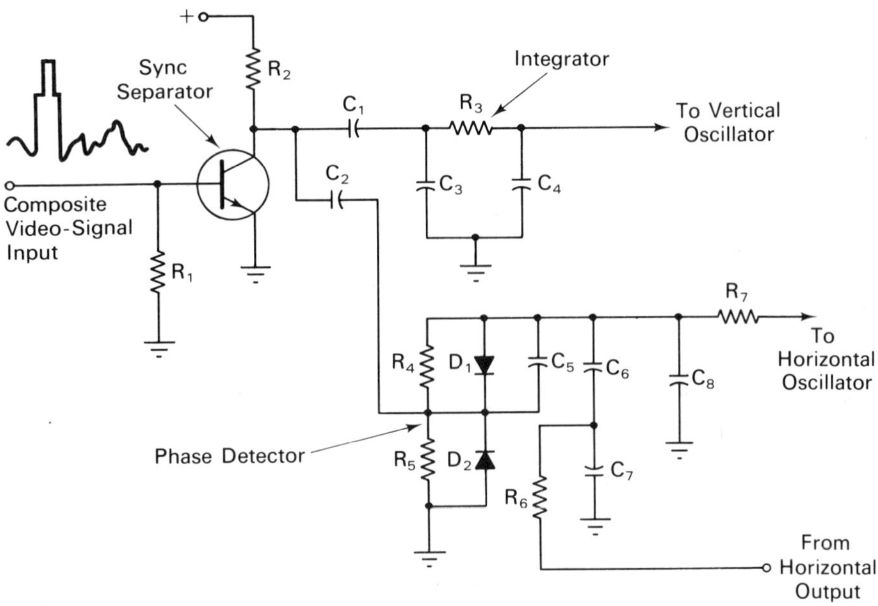

Figure 10-8 Sync separator.

An integrator circuit is used between the separator and the vertical oscillator to process the sync by having successive pulses gradually charge capacitors C_3 and C_4 until the charge amplitude is sufficiently high to lock in the vertical oscillator. The horizontal-rate sync is applied directly to the phase detector diodes as shown for synchronization of the horizontal oscillator.

When sync pulses arrive at the vertical and horizontal oscillators, they must have sufficient amplitude to cause proper vertical and horizontal lock in. Otherwise, picture instability will occur, with consequent weaving, picture tearing or total sync loss as shown in Fig. 1-3. Thus, troubles in the separator affect both vertical and horizontal stability and can be caused by slight changes in the value of components, a defective transistor or insufficient video amplification in the video amplifier.

An oscilloscope provides a handy means for checking the presence of the incoming signal at the base of the transistor (Figures 6-8 and 6-9) as well as at the output of the sync separator (and at the inputs of the vertical and horizontal sweep oscillators). If the scope shows loss of signal amplitude (as indicated on the service notes or schematic of the receiver) other tests must be made, including voltage tests, resistance readings and component testing.

10-7 SUBCARRIER OSCILLATOR AND CONTROL

As mentioned in Sec. 10-2, the color-signal carrier (termed a *subcarrier* to distinguish it from the primary video carrier) is suppressed at the transmitter and only the color sidebands are sent out. Hence, color receivers use a crystal-controlled oscillator to generate the color carrier so it can be re-introduced into a composite signal with the sidebands. Even with crystal control, however, a slight frequency drift can occur (where a phase difference begins to appear between the oscillator and the 3.58 MHz burst) and can cause color contamination. To obtain the rigid control necessary, a phase detector circuit is used in similar fashion to those shown in Figures 10-6 and 10-8. Such a phase detector is linked to a reactance-control transistor as shown in Fig. 10-9. Thus, the phase detector compares the signal obtained from the output of the oscillator (via the buffer amplifier) with the signals obtained from the burst amplifier (the 3.58 MHz burst). If any deviation occurs a correction voltage is generated in the reactance control and applied to the base input circuit and crystal of the oscillator to correct for differences.

As shown, several manual adjustments are provided. The input transformer can be tuned so that the resonant circuit passes a maximum amplitude of the incoming burst signal. Below the phase detector a tint control is shown that may be adjusted for proper flesh tones, ranging from greenish hues to pink and reddish values. Resistor R_8 provides for a manual adjustment of the range through which the reactance control functions. This adjustment permits the reactance control to alter the oscillator frequency both above as well as below the 3.58 MHz subcarrier signal.

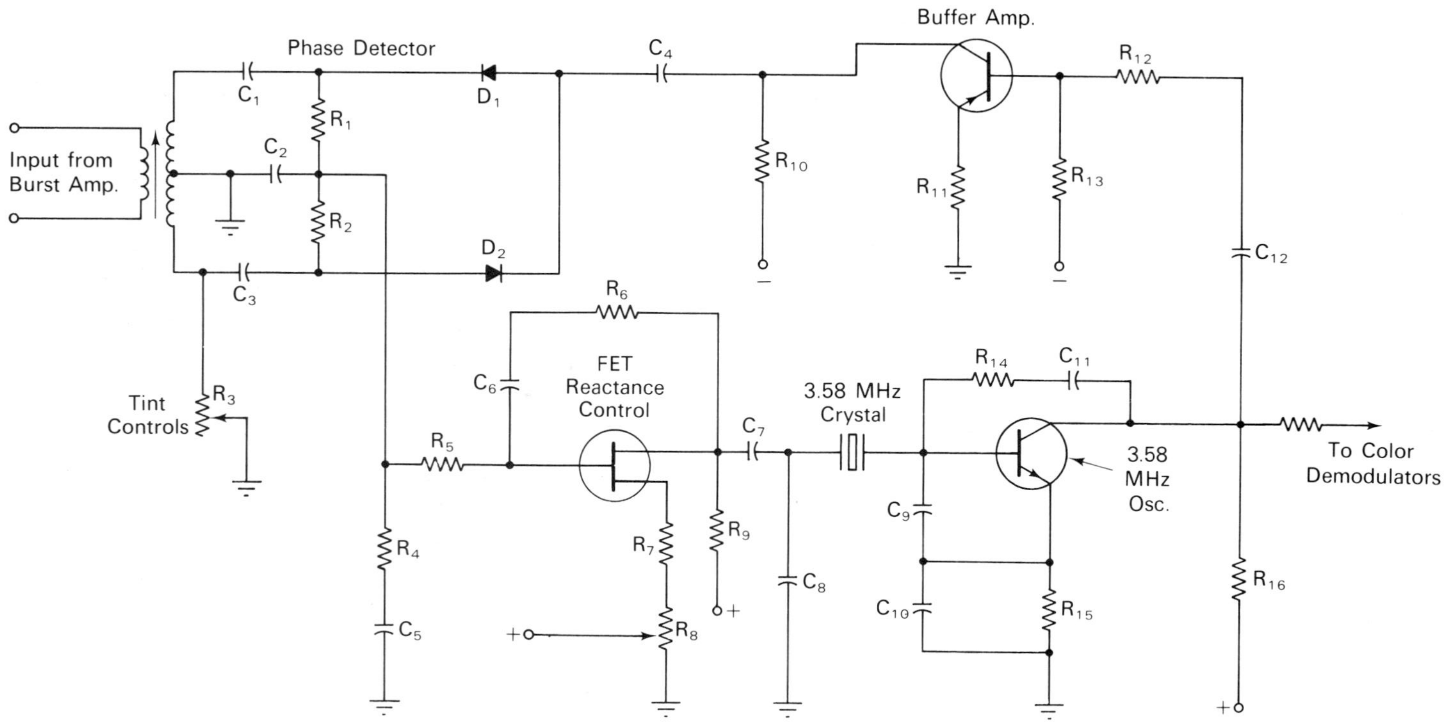

Figure 10-9 Subcarrier oscillator and reactance control.

If the subcarrier oscillator drifts because of lack of reactance control, false colors will appear on the screen and these cannot be changed much with the tint control. Hence, a check must be made of the reactance circuit and transistor, as well as of the burst amplifier and the phase-detector system. Presence and amplitudes of signals can be observed with a scope capable of measuring the 3.58 MHz signal.

Additional checks involve the video-amplifier circuitry, because a faulty operation here will reduce the signals applied to the phase-detector circuits. The oscillator transistor and the buffer amplifier (if used) must also be tested for abnormal conditions in voltage readings, ohmic values of resistors and open or shorted capacitors.

Once the input transformer has been tuned for maximum signal at the phase detector, and R_8 has been set for center 3.58 MHz operation, color synchronization should be maintained without trouble. The viewer's personal preference must be exercised respecting the adjustment of the tint control, since some individuals prefer more or less pink flesh tones. Variations from the selected tint will occur, however, because of differences in the settings of the various cameras used to televise the scenes. A noticeable difference will often be observed between the color intensity and range of commercials compared with live or taped programs. Thus, the tint control should be set for the preferred hues that represent the average for the programs being observed.

10-8 COLOR DEMODULATORS

Blue, red and green representation signals are required in color transmission, but use of three would require special bandwidths. Hence, the three color signals are converted to what are termed *I* and *Q* signals (see Appendix E). At the receiver it is necessary to reestablish the original three primary colors, minus the luminance (Y) signals. This is done by mixing the chroma signals with the subcarrier signal obtained from the 3.58 MHz oscillator and demodulating them to obtain the red minus the luminance (R-Y), B-Y and G-Y.

As shown in Fig. 10-10A, the R-Y and B-Y are obtained from the two demodulators shown, and the G-Y is procured by using a resistive mixing (matrix) circuit. As shown, the Y signal is applied to the picture-tube cathodes, and the R, B and G color signals to the control grids. Some manufacturers arbitrarily designate the demodulators X and Z for particular phase relationships between the two signals (usually 105° apart).

Diode demodulators are shown in Fig. 10-10B, though transistors can also be used (see Fig. 3-11) as well as vacuum tubes. The 3.58 MHz

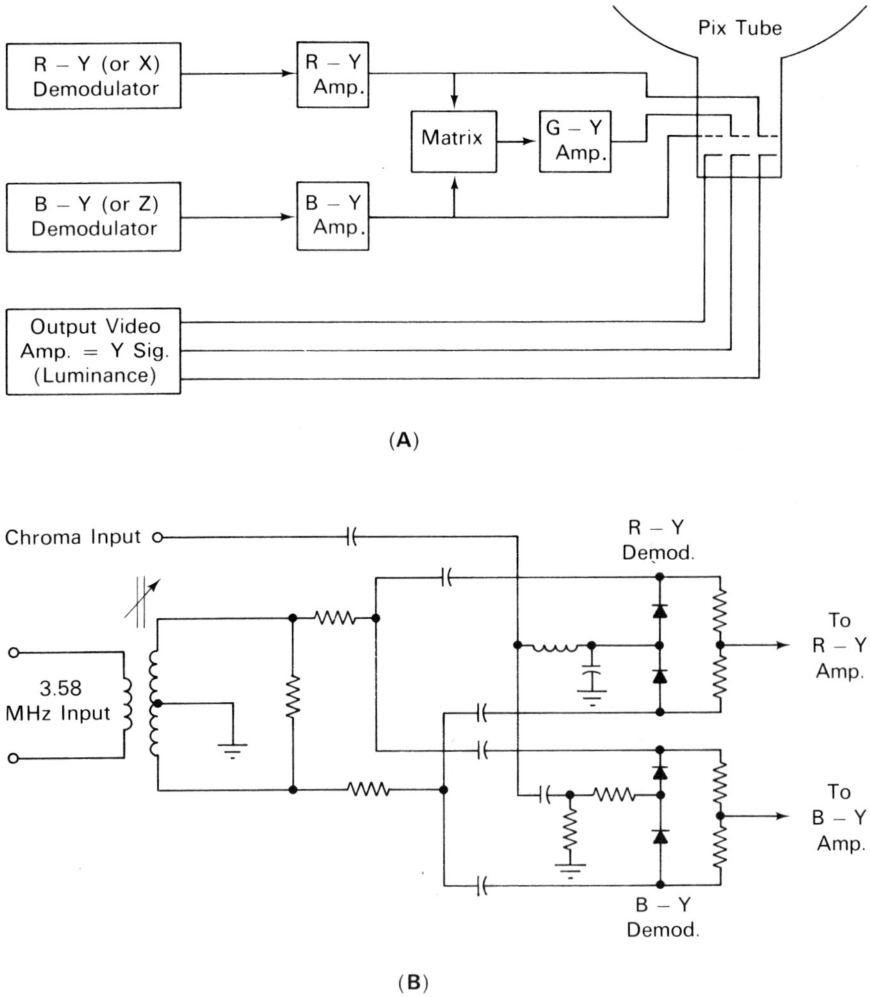

Figure 10-10 Basic color detector system.

subcarrier signal generated by the reactance-controlled oscillator is applied to the input transformer as shown. An adjustable variable core tunes for maximum signal acceptance. The secondary of the input transformer has a grounded center tap, providing for a signal and phase splitting network for application of the 3.58 subcarrier to each demodulator system. The chroma signals are applied to the junction of the detector diodes. Thus, the subcarrier signal is mixed with the chroma and detected at the same time to produce outputs of R-Y and B-Y signals.

If troubles occur in these detector circuits an oscilloscope can be used to ascertain the presence of signals at the secondary of the transformer as well as at each detector diode junction. The peak-to-peak signal voltage at the junction of a pair of diodes may range between 3 to 5 V. The input transformer should be tuned to obtain a maximum signal level. Failure of a demodulator system to produce an output calls for a check of components for off values, shorts or open circuit conditions (see also Sec. 3-10).

10-9 DEFLECTION AND HIGH VOLTAGE

A sweep output system for a black-and-white receiver is shown in Fig. 10-11 (see also Fig. 2-7 for the comparable tube-type version). The vertical output sweep signals are applied directly to the vertical deflection coils. With tube-type sweep systems, the higher impedance output of tubes requires a matching transformer for impedance step down to the coils.

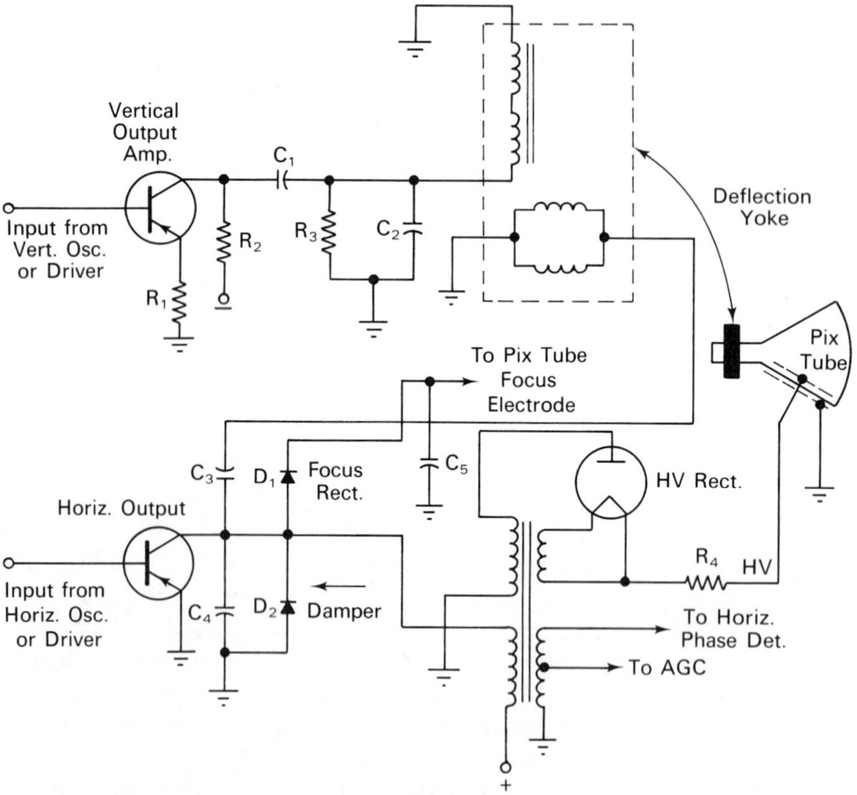

Figure 10-11 Sweep output systems (B/W).

The horizontal output system must perform many functions. Initially it develops the horizontal sweep signals that are applied to the horizontal deflection coils as shown. A damper diode minimizes undesirable transient signals that develop because of the high-voltage pulse signals processed. A diode rectifier supplies several thousand volts for the focus control electrode of the picture tube. Capacitor C_5 is a high-voltage type that filters the ripple component. Failure to obtain good focus calls for a check of the diode as well as a possible short in C_5 (or leakage).

The pulse signals developed in the horizontal output tube are stepped up and applied to the anode of a high-voltage rectifier (either tube type or solid state). The high voltage is applied to the second anode of the picture tube, which consists of a conductive inner coating. The high voltage on this inner coating is applied as shown in Fig. 3-10 and helps accelerate the beam within the picture tube. The outside surface of the picture tube is also coated and is grounded as shown in Fig. 10-11. Thus, a capacitor that is formed between the inner and outer coatings acts as a filter to produce smooth dc. Because of the capacitor function, however, a charge is maintained for a considerable length of time after the receiver is shut off, hence presenting the danger of shock if touched during servicing procedures.

No attempt should be made to measure voltages at the collector of the horizontal output transistor or tube, or meter damage may result. Observe the precautions given in the service notes for the receiver under test. However, a check can be made for presence of lower-voltage pulse signals at transformer windings feeding phase detector or automatic gain control circuitry.

When defects occur in the vertical system, height for picture (or raster) declines. For defects in the horizontal system, however, loss of high voltage may occur and thus cause the screen to go blank (no picture or raster). In many horizontal systems the low voltage is given an additional boost. Consequently, defects in the horizontal output system could also affect audio and IF circuits that are being fed by the boosted low-voltage supply.

Initial checks call for testing of the high-voltage rectifier, horizontal and vertical output transistors (or tubes) and component parts as detailed in earlier sections.

The vertical and horizontal coils are positioned around the tube neck as shown in Fig. 10-12, and together they form the *yoke*. Since magnetic fields deflect electron beams at right angles, the vertical coils are positioned on each side of the tube neck and the horizontal coils at top and bottom. Since both coils are mounted in a common winding-housing (the yoke) any rotation of the yoke tilts the picture. Similarly, if the yoke is not tight against the flare of the picture tube, picture reduction will

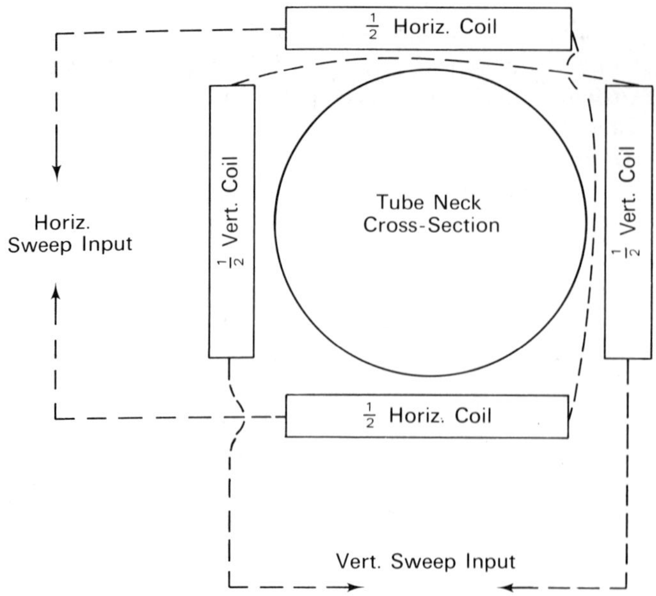

Figure 10-12 Deflection coils (the yoke).

result. If the high voltage declines it will tend to cause the picture to expand (bloom) because of a decrease in beam velocity and hence greater deflection by sweep signals.

The high-voltage and output systems of color receivers perform functions similar to those used in black and white receivers, except for higher voltages required (up to 30 kV in larger-screen sets) and the need for regulation. A transistorized system is shown in Fig. 10-13 and a series stack of silicon high-voltage rectifier diodes is used as shown, with a similar arrangement for the focus voltage. Since the focus output is tapped below the maximum high voltage output, less voltage is obtained. The regulator transistor has a base input variable resistor for adjusting the amount of high voltage. The transistor tends to stabilize the high voltage around the amplitude set by the variable control.

The high voltage must be set for the amount required for proper operation of the color picture tube. Thus, if 25 kV is indicated in the schematic for a particular receiver, that voltage should not be exceeded, otherwise dangerous arcing, component damage and the production of above-average x-rays will result (see Sec. 10-1).

Because the high voltage depends on a horizontal output signal, loss of deflection high voltage could also be caused by faults in circuitry feeding signals to the horizontal output tube. Such circuits consist of the horizontal oscillator and driver as shown in Fig. 10-14. The horizon-

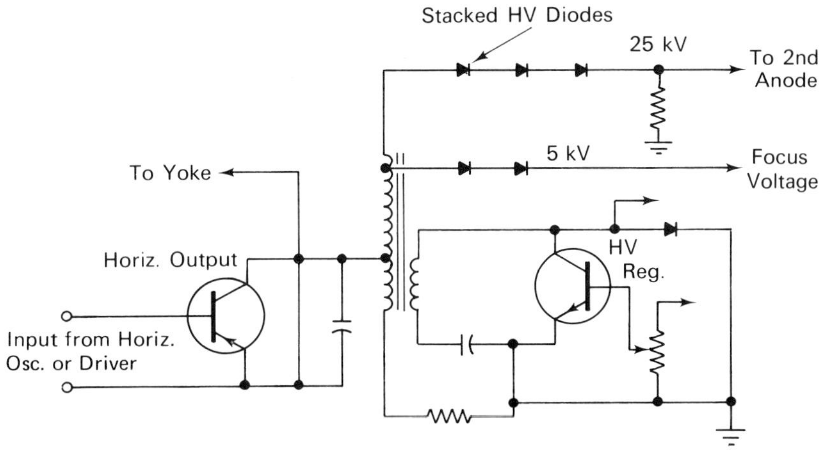

Figure 10-13 HV system in color TV.

tal hold must be adjusted for good stability of horizontal scan to provide proper signals for the driver transistor and the horizontal output. Again, if troubles are evident, both transistors must be checked. If these are satisfactory, components and voltages are tested.

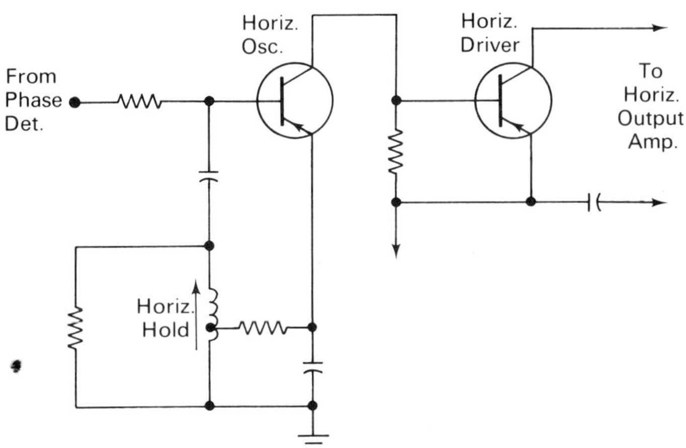

Figure 10-14 Oscillator and driver.

10-10 COLOR PICTURE-TUBE FACTORS

As shown in Fig. 10-15, the luminance (Y) signal is applied to the cathodes, with a variable resistor for the blue drive and another for the green. Note that these cathodes correspond to the control-grid color signal inputs, hence signal drive for a particular color can be regulated. The blue and green settings above and below a given red input (to cathode pin 4) permit obtaining an equal level for each. Thus, decreasing the blue and green drive controls accents the yellow-red color range, and increasing blue and green will diminish (by comparison) the yellow-red color range.

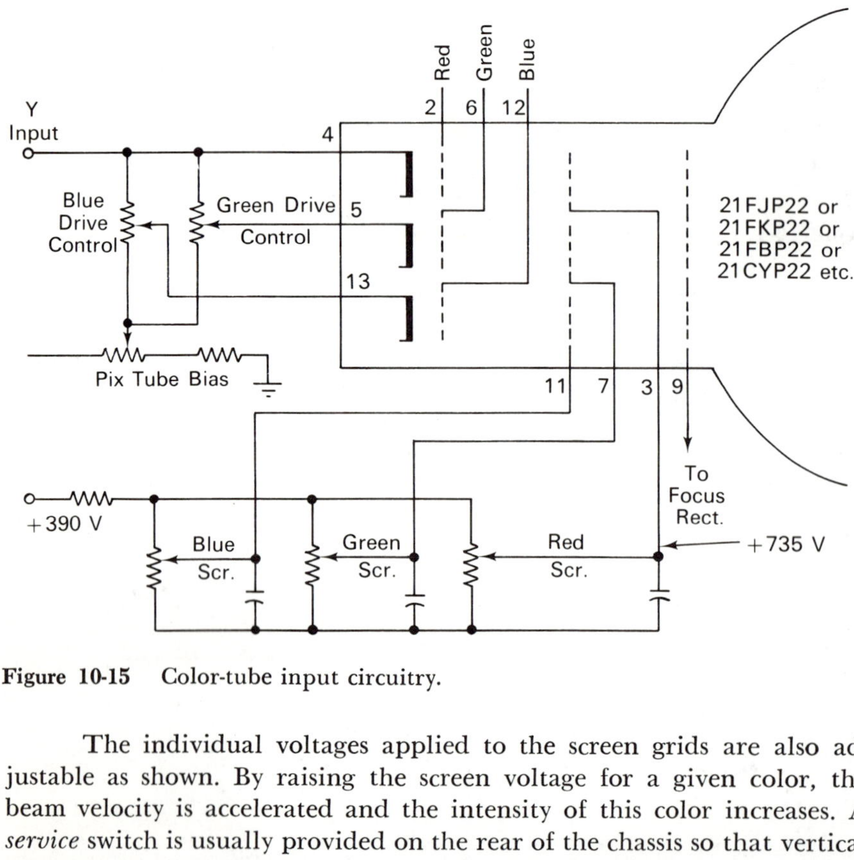

Figure 10-15 Color-tube input circuitry.

The individual voltages applied to the screen grids are also adjustable as shown. By raising the screen voltage for a given color, the beam velocity is accelerated and the intensity of this color increases. A *service* switch is usually provided on the rear of the chassis so that vertical deflection can be reduced to a minimum during adjustment of the screen controls. Two of the screen controls are then turned down (or shunted by 100 kΩ resistors) so only one screen input may be adjusted at a time. If, for instance, the red and green are turned down, the blue is adjusted until a blue horizontal line appears faintly. Now the blue and red controls are bypassed with the resistors and the green screen control adjusted until a green line just barely becomes visible. Next, the green and blue leads are bypassed with the resistors and the red screen control adjusted. The resistors are then removed and the service switch returned to normal.

In Fig. 10-15 the picture-tube bias is regulated by the variable resistor that varies the applied voltage to the cathodes and thus permits changing the overall brilliancy. (Contrast in a color receiver is adjusted with the *color* control, while the *tint* control regulates flesh tones.)

Purity refers to the ability of the tube to produce a uniform field of one color on the screen. Purity is adjusted by loosening the yoke clamps and moving the yoke back toward the base of the neck. The blue and green screen controls are again by-passed by 100 kΩ resistors, thus producing a red field on the screen. The purity rings located on the neck (Fig. 10-16) are then adjusted for a uniform red field. The yoke is then pushed back against the flare of the picture tube. (Normally the blue and green fields will also be pure once the red has been adjusted, provided convergence and other adjustments are all right. Before placing the yoke back, however, the green and blue can be checked.)

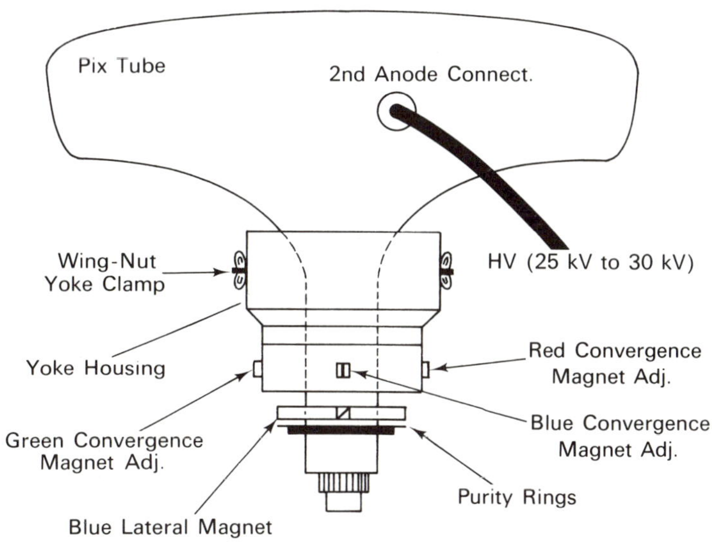

Figure 10-16 Color-tube neck components.

For the three-gun color tube (described in the next section) a special convergence circuitry is included in the horizontal and vertical output system where specific voltages are tapped for application to the convergence coils surrounding the picture tube neck. Adjustments are provided on a special panel in addition to the adjustment set screws shown in Fig. 10-16, as detailed in the next section.

10-11 BASICS OF THREE-GUN COLOR TUBE

The three-gun color picture tube is one of several types available for modern receivers. As the name implies, three electron guns are used, one for processing the electron stream that will produce green colors, another for blue and the third for red. The term *gun* refers to the internal electrodes that process the beam by accelerating it, directing it to the proper segments of the tube face and presenting a sharp focus. Each electron gun has its own cathode circuit, individual control grids and first anode sections.

The phosphor coating on the inside face of this type color picture tube consists of a dot pattern of blue, green and red. These dots produce a colored glow with an intensity that depends on the amount of electrons that strike them. In modern tubes the phosphor dots are surrounded with a black area and also use rare-earth phosphors to produce greater picture brilliance, even in daylight.

In the three-gun picture tube a perforated metal section (termed an *aperture mask*) is situated just before the phosphor-coated inside tube face. The mask has a number of tiny holes that permit entry of the electron beams, as shown in Fig. 10-17. It is essential that each of the three

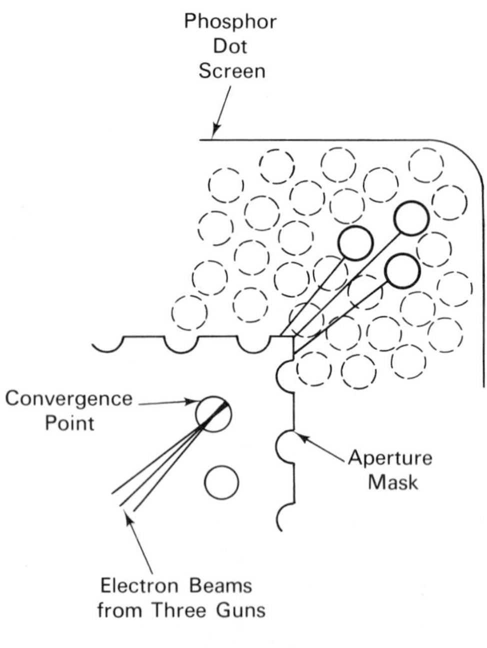

Figure 10-17 Three-gun color tube convergence.

electron beams enters the aperture mask at the right angle so it falls exactly on the color dot corresponding to the particular color it is to represent. That is, the three beams must be adjusted for proper *convergence* on the inner tube face so that proper colors are produced.

To provide for convergence adjustments and to correct for the difference in lengths of travel of the electron beams to the center of the tube and the sides, a convergence-control assembly is located on the neck of the picture tube as shown in Fig. 10-16. This convergence assembly contains several coils that are supplied with signals from the output sweep circuits. These signals influence the electron beams and are essential to achieve good convergence. A typical convergence yoke of this type is shown in Fig. 10-18, where U-shaped ferrite cores are used. A cylindrical magnet plus an adjusting knob is situated within the core gap of each coil. The assembly is positioned on the tube neck so each coil is immediately above the corresponding pole piece to influence the individual beams.

Static convergence adjustments are made by turning the knobs of the convergence yoke and by adjusting the lateral blue-beam-positioning magnet. Dynamic convergence is adjusted by turning special controls that regulate the currents flowing through the coils. Basic adjustments are covered in the next section.

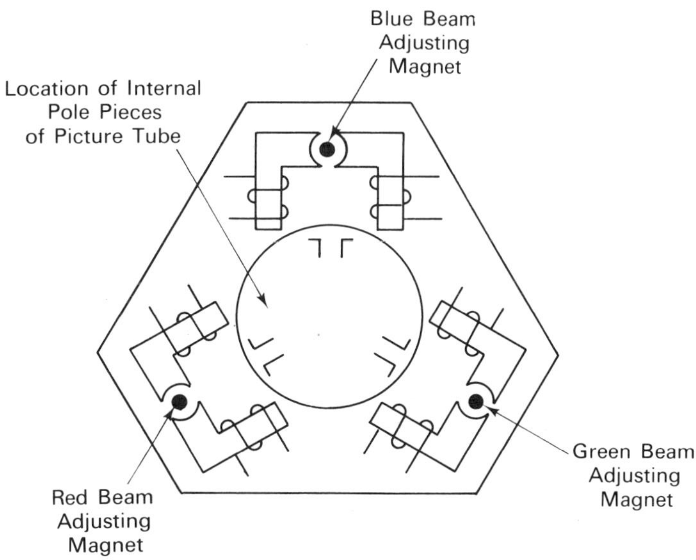

Figure 10-18 Color TV convergence yoke.

10-12 CONVERGENCE ADJUSTMENT FACTORS

White cannot be produced on a color-picture screen unless proper amplitudes of red, blue and green signals are applied simultaneously to their respective phosphor dots of red, blue and green. Thus, the ability of the color receiver to produce white can be used to judge proper convergence. An accurate check of convergence at the center top and sides can be made by employing a generator that produces white dots on the screen (Fig. 10-19).

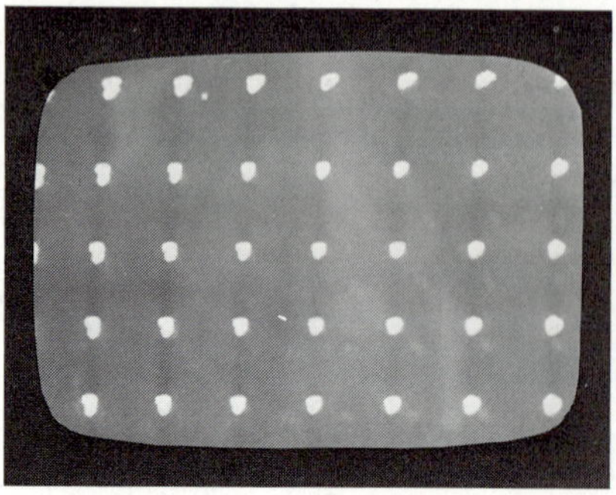

Figure 10-19 Dot pattern.

Note the white-dot distortion at the sides, compared to the more nearly round patterns at the center of the screen in Fig. 10-19. The overlapping of the red, blue and green signals would result in a perfectly round dot and any distortion indicates lack of convergence. A convergence adjustment panel, as shown in Fig. 10-20, is used for setting convergence for the various colors at top, bottom, right and left areas of the screen. Such a panel is usually supplied with a long harness-wiring cable so it can be fastened to the outside rear of the cabinet, but facing front for easy accessibility during convergence adjustments.

Usually, once convergence has been properly set, there is no need for undertaking complete reconvergence unless a new picture tube is installed. Some touch up may be tried during troubleshooting procedures to compensate for aging components. For complete convergence, the center must be converged (static) by convergence-yoke adjustments, after which individual adjustments are made for proper side convergence as well as top and bottom. If the service notes for the receiver are available, the suggested sequence of adjustments should be made. Otherwise, the general factors relating to convergence adjustments as found in the dot generator manual should be followed. Usually the dot generators are combined with the cross-hatch and color generators described more fully in Chapter 11.

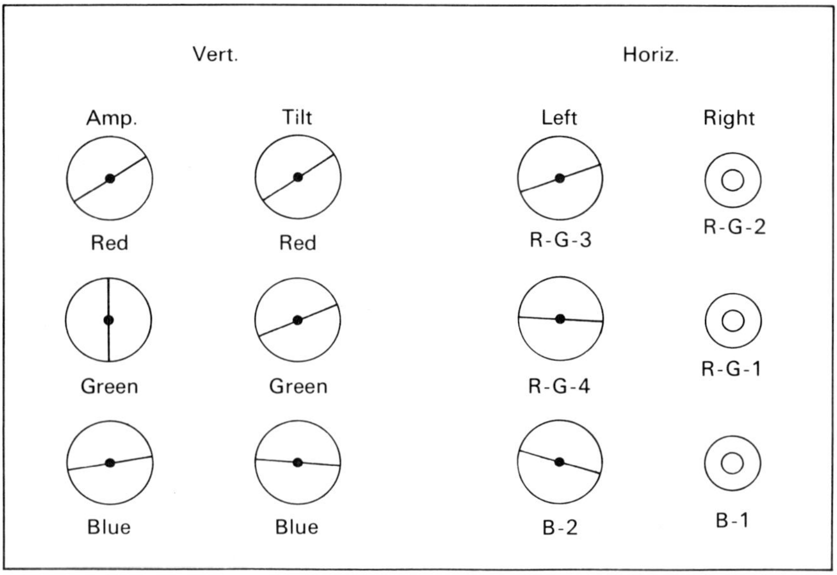

Figure 10-20 Convergence adjustment panel.

On older receivers periodic demagnetizing (degaussing) was necessary to rid the picture tube of residual magnetism picked up by stray fields from vacuum cleaners, hand tools and so forth. A cylindrical degaussing magnet (available from electronic parts jobbers) was energized and passed across the picture-tube face and withdrawn to a distance of about six feet before turning off the device. (If the degaussing magnet were turned off near the picture tube, the collapsing magnetic fields would energize and magnetize picture-tube sections instead of demagnetizing them.) Modern receivers use self-degaussing coils placed around the picture tube rim. Such coils are energized each time the receiver is in use and thus keep sections degaussed. If good purity cannot be achieved, however, a complete demagnetizing procedure should still be tried. (See Sec. 12-24 and Fig. 12-15.)

10-13 TRINITRON COLOR PICTURE TUBE

The Trinitron single-gun picture tube is a development of the Sony Company and the basic principles are shown in Fig. 10-21. Even though a single-gun structure is used, three electron beams are still present. The phosphor screen is in the form of narrow vertical stripes as shown, and the aperture mask consists of an aperture grille made up of

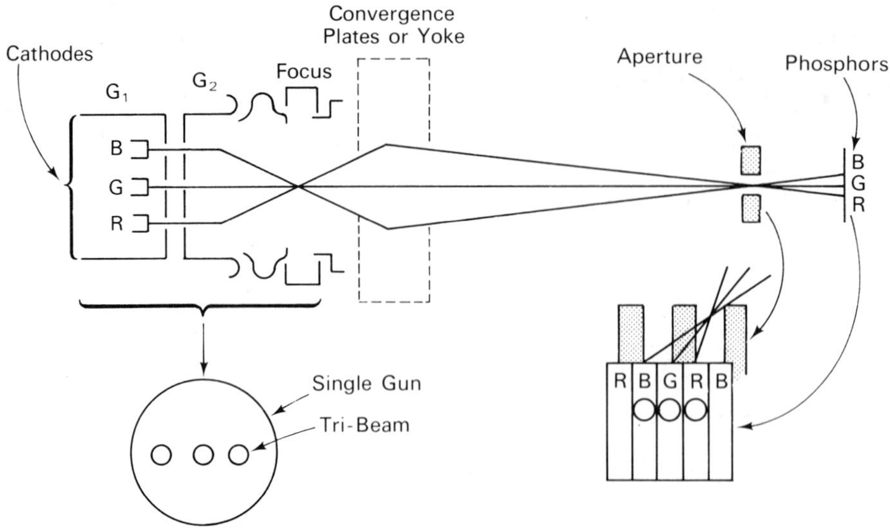

Figure 10-21 Sony's Trinitron color-tube system.

vertical slots through which the beams are aimed at their respective strips of red, blue and green phosphors.

By the use of internal convergence plates (on the smaller tubes) or a special convergence yoke on the larger tubes, a simplified convergence method is achieved. Thus, because of the single-gun design, the vertical dot structure and aperture grille, many of the convergence problems of the three-gun tube are eliminated. Hence, convergence may require only a few adjustments and are simple.

For this tube conventional circuitry is used for applying the various signals (chroma to the grids and the luminous signals to the cathode). Variations in circuitry occur with all receivers, however, and the schematic for a particular type should be consulted when servicing.

10-14 IN-LINE COLOR PICTURE TUBE

Another color picture tube of different design than the three-gun type that was in exclusive use for so many years is the in-line tube developed by RCA. As shown in Fig. 10-22, a vertically-striped phosphor screen is used as with the Trinitron, though the aperture mask is a slot-type affair as shown. This design essentially provides a cross-tie between apertures and makes for a more rigid mask structure. A three-gun assembly still is employed, but the three guns are now aligned side by side as shown.

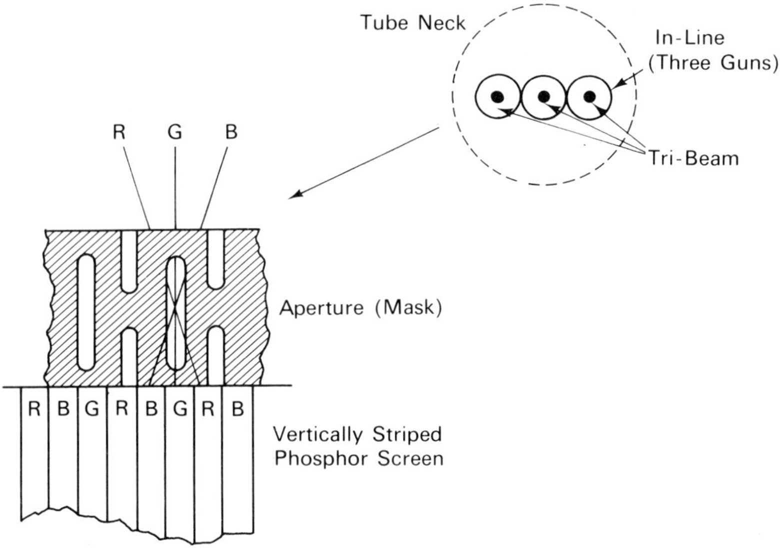

Figure 10-22 RCA's In-line color tube system.

A built-in self-convergence feature is provided with this in-line tube. A toroid deflection yoke is used that has a precision-wound coil core and is permanently attached to the tube neck at the factory. Consequently, prior to cementing the yoke to the tube, it is properly aligned and hence eliminates the need for any but minor adjustments in the home. As dynamic convergence magnets are no longer needed, the convergence procedures are eliminated.

The RCA color tube in-line gun structure is only about half the size of the conventional three-gun (delta-aligned) structure, thus a small tube neck can be used as well as a shorter tube. The yoke also is not as bulky and the horizontal and vertical deflection-coil wires are positioned accurately in preset grooves at the factory, again assuring a minimum of off-adjustment occurrences after the receiver is put into use.

11

television servicing factors

11-1 INITIAL CONSIDERATIONS

In servicing and troubleshooting television receivers initial considerations must always be given to whether the faults that appear are caused by circuit defects or by misadjustment of controls. If the controls have been properly adjusted and the receiver has been functioning properly, a sudden loss of picture or sound, or other troubles such as noise and interference lines, in all probability are caused by component defects that suddenly occur, or by outside interference. However, other stations should be tuned in to make sure the faults are not with a particular broadcast from a specific station.

Components may change values slightly as they age and thus upset control settings. In such instances readjustments are required and the procedures detailed in this chapter (and the preceding one) should be followed. It is unlikely, however, that such gradual aging of parts would cause an abrupt change in the picture or sound. Checks of the controls should be made routinely before attempting the more complex troubleshooting procedures involving measurements on chassis sections.

The precautions detailed in Sec. 10-1 should be kept in mind during any servicing procedures on television receivers. The fundamental factors detailed in Chapter 1 should also be kept in mind and reviewed if not thoroughly remembered.

11-2 SPECIALIZED TEST EQUIPMENT

Vertical or horizontal bars can be produced on television screens for test purposes by using signal generators that inject audio tones into video amplifiers. An RF signal generator modulated by an audio tone between 60 and 15,750 Hz can be applied to the antenna terminals of a TV receiver for producing horizontal bars. If the modulating tone is above 15,750 Hz, vertical bars will be produced.

A convenient test instrument is one that has the ability to produce horizontal, vertical or both type bars simultaneously as required. In addition, instruments are available that will also produce the dot pattern for convergence adjustments, as well as color bars so that phase relationships between the several color signals can be checked and amplitudes tested.

A typical color generator panel is shown in Fig. 11-1. The pattern selector switch permits production of vertical bars, horizontal bars, both vertical and horizontal (termed crosshatch), color bars, dots and a single dot for aid in precise convergence adjustments. The dot size can be adjusted as needed by a variable control as shown. Other controls may include a color amplitude adjusting knob, a dot position knob and a pattern adjustment for better lock-in. The television receiver is usually set to a specific station (without antenna attached) when using the generator.

The sequence of color bars that appear on the television screen is shown in Fig. 11-2. Such color bars are also referred to as a rainbow display and they are produced by phase shifting the subcarrier signal in relation to the burst signal of 3.58 MHz. A keying oscillator is also part of the instrument's circuitry and is used to blank out portions of the rainbow display to produce distinct segments instead of a continuous color pattern.

As detailed in Chapter 4 and illustrated in Figures 4-4 through 4-7, Lissajou patterns may be displayed on an oscilloscope by injecting signals into both the vertical and horizontal (sweep) inputs of a scope. When this principle is used in conjunction with a keyed rainbow color generator an instrument known as a *vectorscope* is formed.

The vectorscope is essentially a well-designed oscilloscope with a special faceplate as shown in Fig. 11-3. Some have numbered segments corresponding to the ten color bars of Fig. 11-2; others may use segments

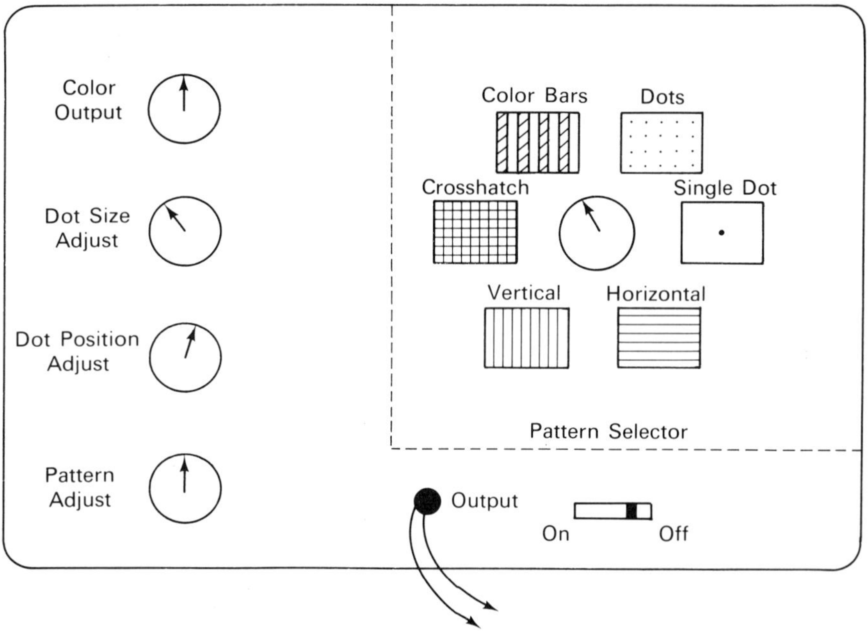

Figure 11-1 Typical color generator panel.

Figure 11-2 Color bar pattern.

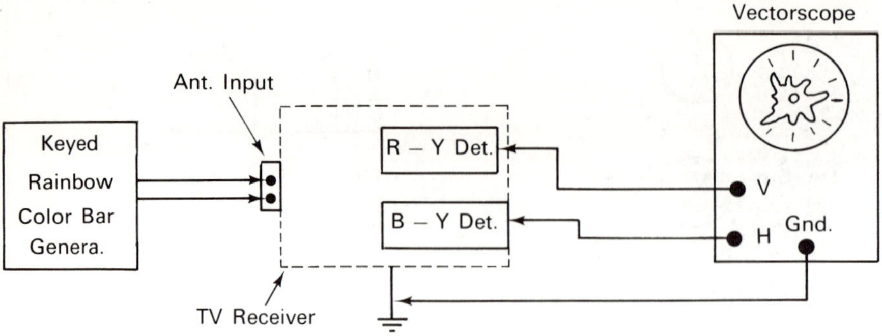

Figure 11-3 Basic vectorscope test connections.

identified by their respective vector angles. Some commercial types have the rainbow generator built in.

As shown in Fig. 11-3, the R-Y signal from the color demodulator is applied to the vertical input of the vectorscope while the B-Y signal is injected into the horizontal for sweep purposes. With the color bar generator output applied to the antenna input terminals as shown, a vectorgram petal formation is created as illustrated in Fig. 11-4. The phase relationships between the signals is shown in Fig. 11-4A with the petal formation. The pattern appearing on the vectorscope screen is

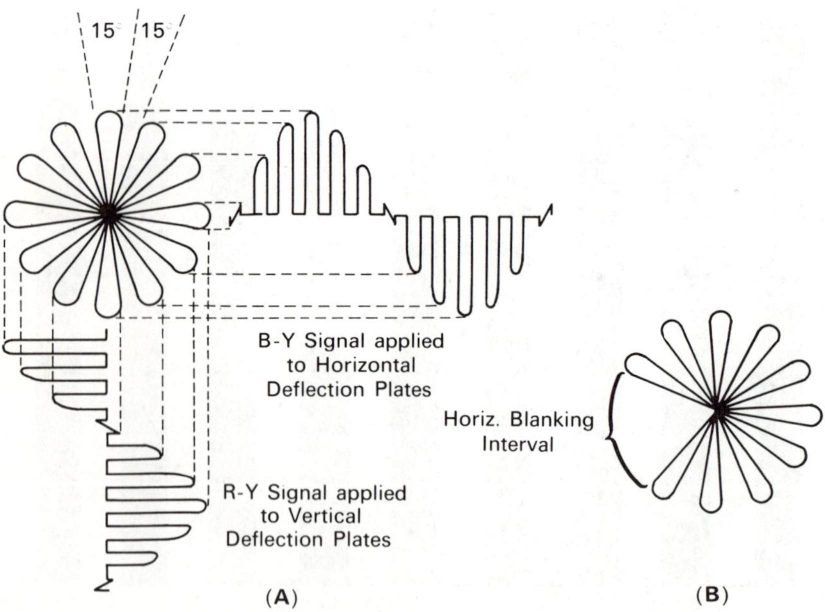

Figure 11-4 Vectrogram petal formation.

shown in Fig. 11-4B. This is the ideal pattern, though the actual one is somewhat distorted as shown on the screen in Fig. 11-3.

When some of the displayed vectorscope petals are longer than others, nonlinear distortion is indicated. With flattened petal tops, some circuit overloading is present in the receiver (improperly set AGC [see Sec. 3-11], too high a color-control setting or off-value components.) Many other checks are possible with the vectorscope, such as location of defective color stages, mistuned bandpass amplifiers, incorrect subcarrier oscillator tuning and inoperative color amplifier or other color circuits.

The inoperative circuitry (loss of red chroma signals, for instance) will collapse the petal display into a horizontal line. Similarly, loss of the blue (B-Y) signal produces a vertical line, since the horizontal sweep formed by the B-Y is missing.

Wide petals that overlap indicate chroma circuit troubles or misalignment of IF, bandpass or subcarrier circuitry. If the petal formation is blurry in appearance, trap adjustments (3.58 MHz) may be needed. Additional data on usage will be found in the manual for the vectorscope, especially as it relates to that particular model.

If a separate color bar generator is used with a scope, the bar generator must be operating properly for best results because it is necessary to produce well-defined sections or the vectorgram pattern distorts. The combination of generator and scope can be tried on a receiver known to be in good operating condition. Thus, the vectorgrams appearing on the scope can be evaluated regarding their usefulness.

11-3 LINEARITY DEFECTS

If a picture is too wide or has insufficient width, the width control must be adjusted so the picture just fills the screen. If a drive control is present, it too may have to be adjusted. The relationship of sweep amplitude to picture width is shown in Fig. 11-5A. These same factors apply to the vertical sweep regarding insufficient height or excessive height.

Note that the sawtooth waveforms in Fig. 11-5A all have a linear rise—that is, a gradual incline. When a sawtooth is distorted it will alter the sweep linearity as shown in Fig. 11-5B. If the distortion manifests itself as poor lows with a flat top near the end of the sweep, it will result in a crowded picture at the right because the beam trace slows down and hence does not spread out the video picture sufficiently. If the sawtooth signal has poor high-frequency components, the crowding occurs at the initial sweep at the left of the screen. Again, these factors also apply to vertical sweep (which begins at the top of the screen and sweeps down-

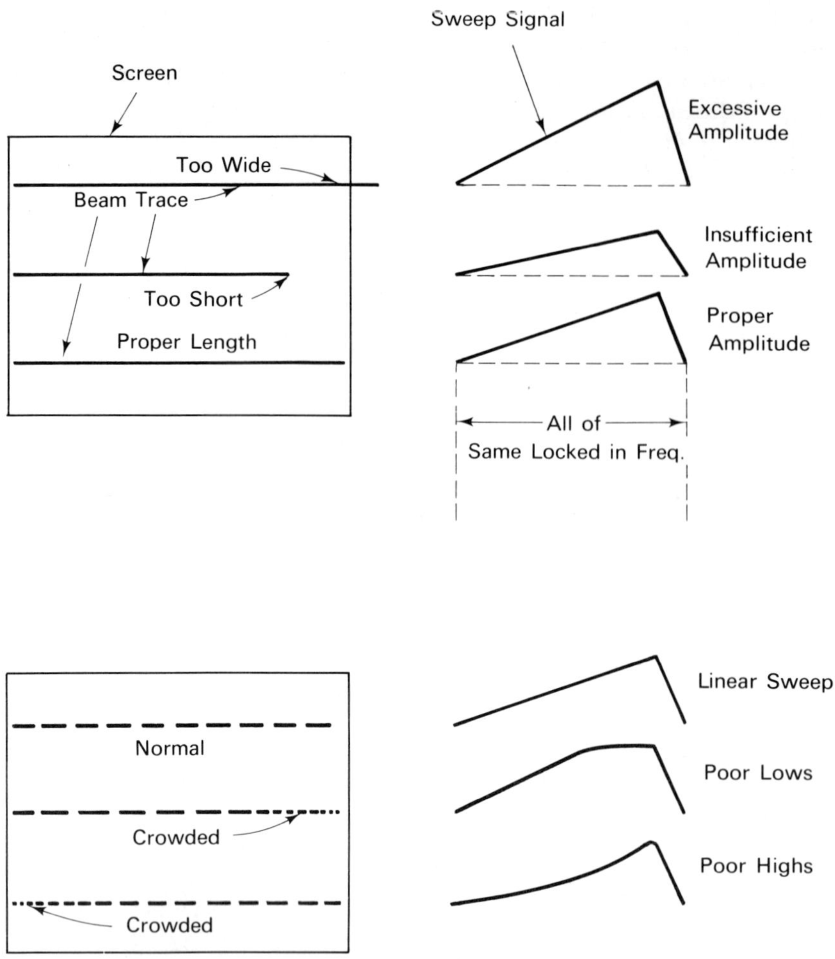

Figure 11-5 Sweep defects in TV.

ward). Thus, for a sawtooth with poor lows, the crowding would occur near the bottom of the screen.

As shown in Fig. 11-6, the bar generator horizontal-line pattern checks for vertical linearity. If there is crowding at either top or bottom, the vertical linearity control must be adjusted. The procedure also involves readjustment of the height control to compensate for the height loss (or increase) that may occur during linearity adjustment. When equally spaced bars are obtained without too high or too short a picture, good linearity has been achieved.

Figure 11-6 Horizontal bars check vertical linearity.

As shown in Fig. 11-7 the vertical-bar pattern is used to check for horizontal linearity. A crowding of bars at either left or right calls for adjustments of horizontal linearity and height controls until an evenly spaced bar pattern is obtained, without overextending the picture or reducing its width from either side. Failure of the vertical and horizontal controls to achieve good linearity calls for a check of the sweep circuitry, particularly the output stages.

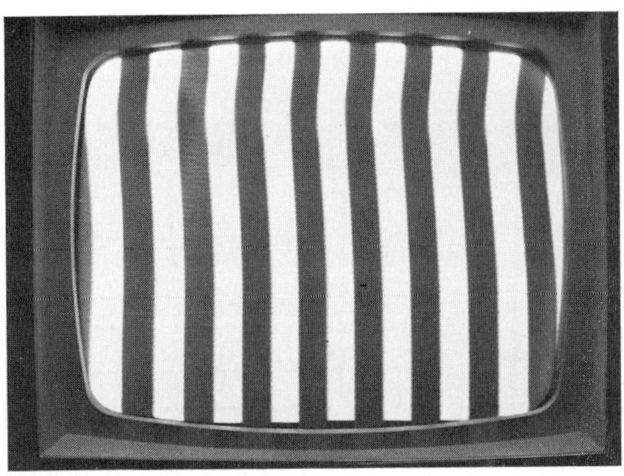

Figure 11-7 Vertical bars check horizontal linearity.

With a cross-hatch pattern as shown in Fig. 11-8, both the vertical and horizontal linearity can be checked simultaneously. In some generators the widths of the bars can be altered to suit. Note, in Fig 11-8, the vertical bars are adjusted to be thicker than the horizontal bars. Such adjustments to the visual pattern depend on the technician's preference and inclinations in the type pattern and procedures best suited to him. Many technicians may take an overall look at total linearity with the cross-hatch pattern, then select either the horizontal or vertical bars alone for making specific adjustments.

Figure 11-8 Cross-hatch pattern checks overall linearity.

The crosshatch pattern will also display curvature defects known as *pincushioning,* as shown in Fig. 11-9. In black-and-white sets pincushioning magnets are sometimes used and are suspended beside the bell of the picture tube by rods or wire supports. To adjust for pincush-

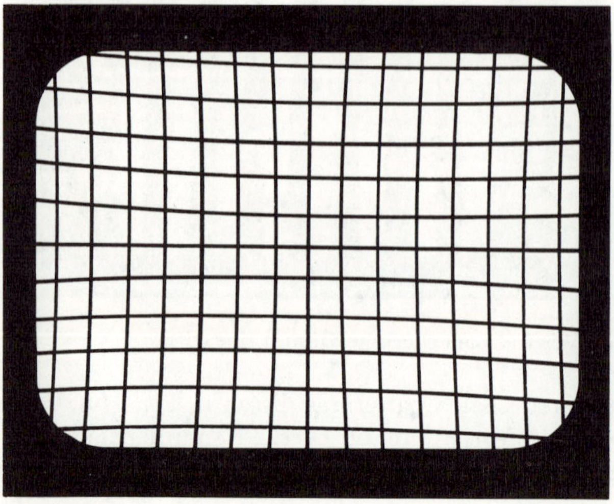

Figure 11-9 Pincushion defects.

ioning, these magnets are positioned closer or farther away from the tube, or are moved around slightly. If pincushioning is too far off the yoke can be pulled back somewhat from the flare of the picture tube. This reduces the raster so adjustments are more easily viewed on the screen. Pincushioning that is severe and cannot be corrected may be caused by shorted turns on the yoke, or defective capacitors or resistors that may be shunting the vertical or horizontal coils.

For color receivers special pincushioning-correction circuits (inductors and resistors) are used. The pincushion transformers are adjusted by turning the tunable core slugs to correct for bending in the vertical and horizontal line structure of the crosshatch pattern.

11-4 COMBINATION DEFECTS

Picture quality and distortion troubles may be caused by a combination of defects and misadjustments. Such a combination of faults is illustrated in Fig. 11-10. Here the vertical linearity needs adjustment (or the vertical circuits require servicing). In addition, the contrast control (or contrast circuitry) requires attention as well as the focus control (or focus circuitry). These defects would also appear during normal picture viewing, of course, though a pattern such as that in Fig. 11-10 or a crosshatch pattern emphasizes the particular faults more readily for identification.

Figure 11-10 Defects in linearity, contrast and focus.

The picture shown in Fig. 11-11 would indicate defects in contrast circuitry if the contrast cannot be brought to normal with the control. If the contrast circuitry (as well as the AGC circuits) is operating satisfacto-

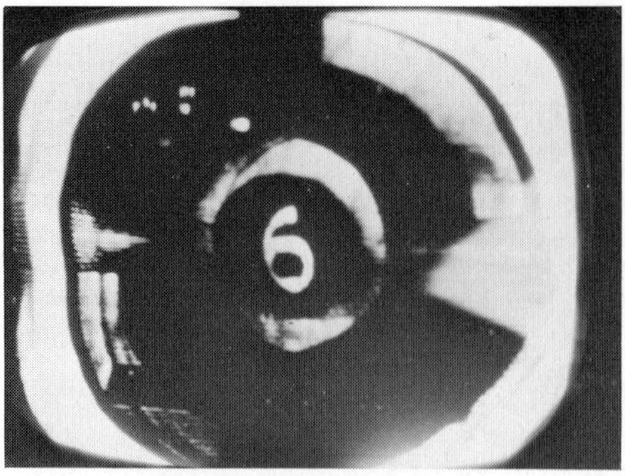

Figure 11-11 Defects in contrast circuitry or pix tube.

rily, the picture tube may be faulty (see Sec. 3-11). The picture tube can be tested with a special instrument described in Sec. 8-6 and illustrated in Fig. 8-13.

A quick check can be made of vertical linearity without test equipment by misadjusting the vertical hold control so that the picture rolls slowly. When this is done the blanking bar also is visible, as shown in Fig. 11-12. This blanking bar should have the same thickness whether it appears at the bottom or top of the screen. If the bar changes in thickness during its vertical travel it indicates improper vertical linearity.

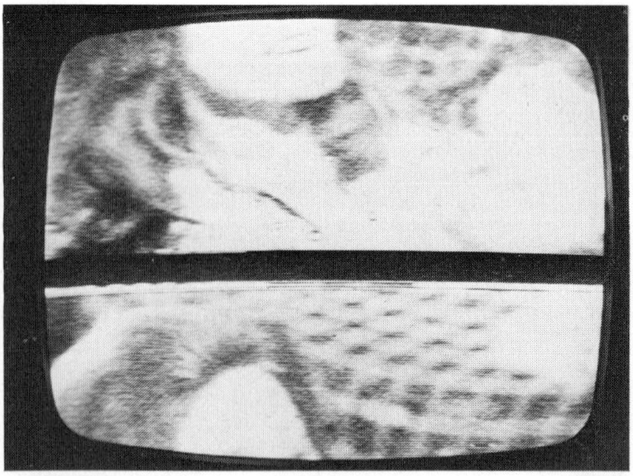

Figure 11-12 Blanking bar checks linearity.

The oscilloscope is useful in checking the type of sawtooth-signal defect that is altering the linearity. The signal displayed in Fig. 11-13 has good (but not perfect) linearity and such a signal will provide a sweep that produces little distortion. Thus the resulting linearity is good enough

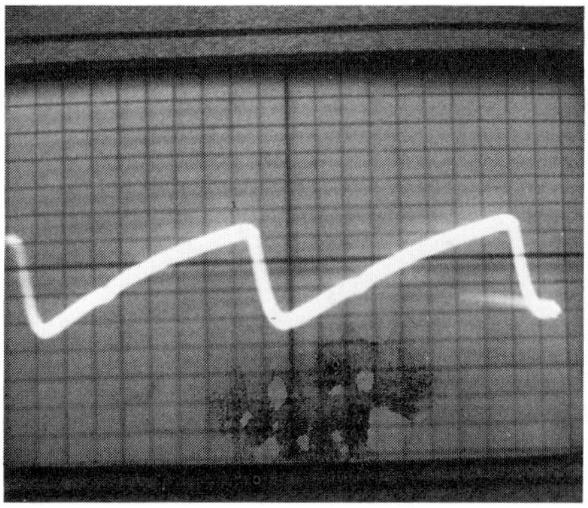

Figure 11-13 Sawtooth signal.

to be tolerated because it would not provide sufficient picture distortion to be noticeable in average scenes.

A severely distorted sawtooth signal is shown in Fig. 11-14, where a dip occurs almost half way up the incline of the sawtooth. Such a dip would cause a slowing down of the sweep near the middle of the screen and a crowding effect would occur. If the dip were deep enough the crowding would be so close that a bar-effect would be noticeable on the screen.

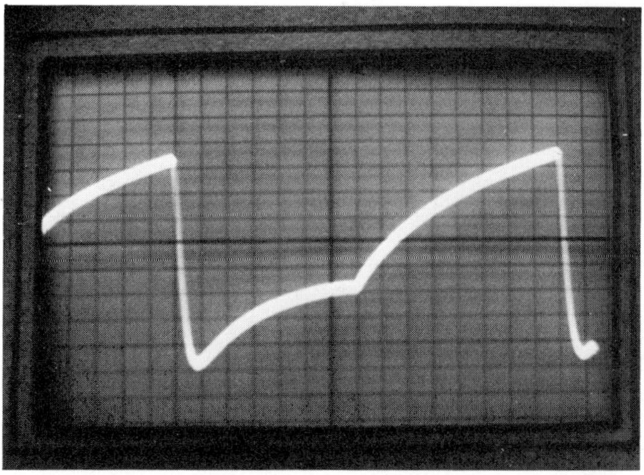

Figure 11-14 Sawtooth with distortion.

The sawtooth voltages can be checked throughout the vertical or horizontal oscillator circuits, though caution must be observed in the output sweep amplifiers because of signal magnitudes. *In most receivers it is not advisable to take any measurements, whether with a scope or voltmeter, at the output of the horizontal amplifier transistor (or tube)* (see also Sec. 10-1).

12

representative troubleshooting examples

12-1 INTRODUCTORY NOTE

When electronic equipment does not function properly a number of approaches may be taken to diagnose the symptoms, make tests to localize the particular circuit involved and, finally, pinpoint the defective component or parts causing the trouble. Of course, when the problems are simple and readily evident not much testing is required. Thus, a loose plug in the connecting cable of a hi-fi system may be found at once and sound restored, or a momentary power surge may have tripped a circuit breaker that only requires resetting. In such an instance, however, it may be advisable to make some checks on power consumption or currents drawn from the line to make certain that no partial short exists. The more complex the fault, however, the greater need for using logical test methods and taking advantage of the time-saving features of test equipment.

Since various approaches may be used in testing and troubleshooting procedures, the examples given in this chapter are not intended to illustrate an inflexible course for the symptoms indicated, but rather to demonstrate the logic that should be followed and the manner in which the principles detailed in earlier chapters are followed. Uses and applications of specific test equipment in the examples are limited to the particular needs for a particular troubleshooting procedure. Hence, all the possible uses of such equipment are not covered in the examples and reference should be made to earlier chapters whenever a review of extended applications is needed. Thus, an ohmmeter may be used to check whether a fuse is open without resorting to other uses for an ohmmeter, such as signal tracing, reading resistance values and so forth.

In the examples given, it is assumed that essential basic equipment is on hand. Without certain equipment, testing and troubleshooting procedures will still localize the problems, though the processes will take longer. If, for instance, an ammeter (or milliameter) is not available, the voltage across a known-value resistor can be measured and current calculated on the basis of E/R.

12-2 CASE-HISTORY INDEX

Since a particular trouble-shooting example given in this chapter might parallel a specific problem encountered by the reader, this section lists the topics covered. This index expedites identifying a certain testing and servicing procedure that may be useful and applicable in a similar situation encountered again.

Hi-Fi (AM, FM, Stereo) Case Histories

Television Case Histories

12-3 INTERMITTENT CHANNEL IN STEREO PHONO

A low-priced stereo phonograph amplifier had an intermittent left channel. The sound output would crackle and become noisy at times, with a decrease in volume. The condition would stop only for a few minutes and then return.

Troubles that occur infrequently present problems in servicing because all checks are normal while the unit is operating properly between intermittent malfunctions. For this unit, however, the intermittent condition was frequent enough to make testing easier. Initially the speakers were interchanged, but the left channel still exhibited the intermittent condition and the right channel sound was normal.

The input leads from the phonograph pick-up were also reversed to make certain that the phono unit was not at fault. With the phono input cable reversed, however, the left channel still showed intermittent operation, indicating the fault was in the circuitry. The chassis was removed from the cabinet and the unit turned on. A constant-amplitude signal was injected into the left channel input from a signal generator. A scope was applied to the collector circuit of each stage and left there for a few minutes to check for an intermittent condition. All circuits prior to the output transformer checked out satisfactorily. A scope picture obtained at test point 2 in Fig. 12-1 for the base of Q_2 exhibited an intermittent condition, while the pattern at test point 1 did not. A check at the base of Q_3 also indicated an intermittent signal, but not as great a change occurred here as at the base of Q_2. At test point 3 the intermittent condition was greatest.

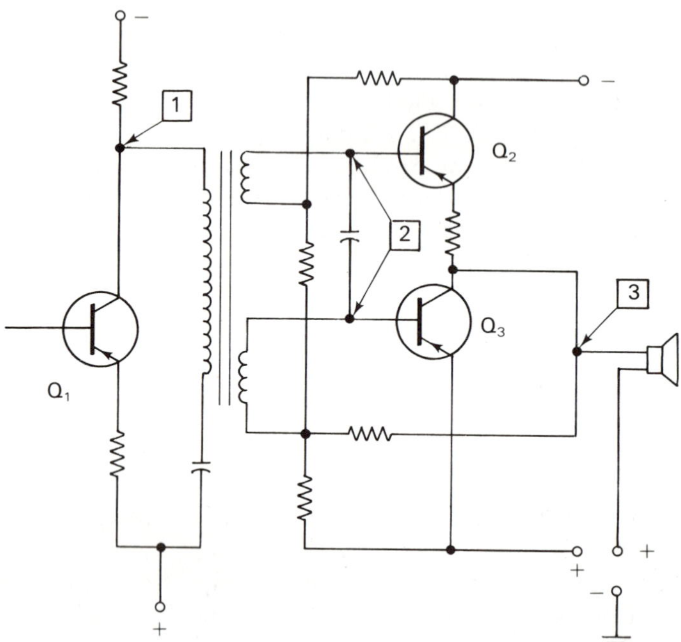

Figure 12-1 Checks on intermittent output amplifier.

If the intermittent condition began within the transistors, there would not be too much of a change at the base input unless some intermittent short was the cause. However, since faulty transistors could produce such symptoms at both base and collector circuitry, the transistors were removed and tested. Both were found to be normal. While these transistors were out of the circuit, scope patterns were observed at both the upper secondary winding of the transformer and the bottom. The intermittent condition appeared more pronounced in the upper reading than the lower. Since the signal was intact at test point 1, the checks showed a defective output transformer. Upon replacement the intermittent condition was no longer present and proper operation was restored.

12-4 POOR GROOVE TRACKING OF PHONO NEEDLE

The phono cartridge and arm assembly of a record player did not track properly in the record grooves and would tend to slide over the record. A new diamond needle had been installed only two weeks ago.

An inspection of the needle assembly showed severe lint clogging around the needle point. The lint was so impacted that a camel-hair brush was ineffective in removing it until a cleaner fluid was used to soften it. The turntable was also found to tilt slightly and a small carpenter's level was used to make the correction. The unit then performed satisfactorily on all records at hand.

12-5 INTERMITTENT RECORDING IN REEL-TO-REEL TAPE MACHINE

A reel-to-reel tape stereo recorder with automatic reverse performed intermittently in the left channel in the forward mode. On reverse play performance was satisfactory. A test tape was played and the tape recorder performed satisfactorily, indicating the trouble was caused by faulty recording. Since the circuits were in use for both recording and playback (except the bias oscillator) the possible causes were pinpointed to the heads, bias circuitry or switching systems.

As a matter of routine servicing, both heads were cleaned and demagnetized. Each head had inputs as shown in Fig. 12-2A. With the machine switched to the recording mode and in the forward tape direction, a sawtooth signal was injected into the microphone inputs. Next a scope was used to check the signals arriving at the recording head as shown in Fig. 12-2B. At test points 1 and 2 the sawtooth signal was observed, while at points 4 and 5 the sinewave signal generated by the bias oscillator appeared. All signals were steady and showed no variations.

Because an intermittent condition can disappear for a while, the scope was left on each terminal for a few minutes. Also, the switches were checked to see if any poor contacts were made. Often, in automatic reverse recorders, the numerous switching contacts contribute to problems. All contacts appeared normal, however, though a contact cleaning fluid was used to assure good electrical connections. A check on performance, however, indicated the recording still became intermittent on occasion in

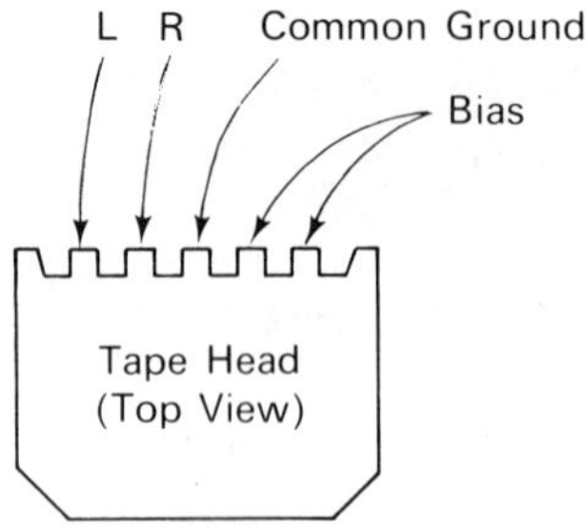

(A)

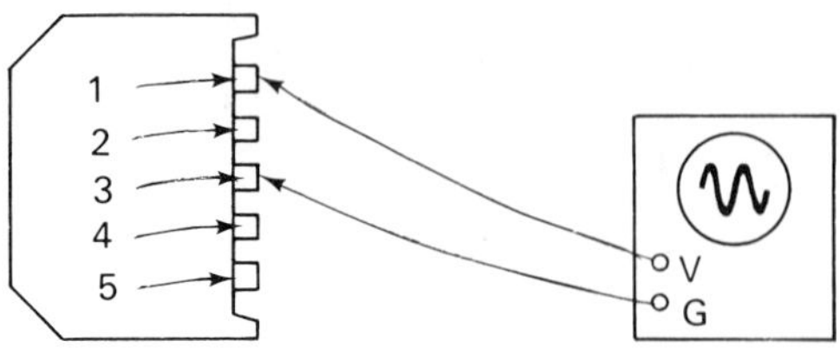

(B)

Figure 12-2 Scope checks on tape head.

the left channel. Thus, it appeared the fault was either in the recording head or the soldered connections to that head. The L and R terminal wires (shown in Fig. 12-2A) were unsoldered and reversed to see if the intermittent condition then changed to the right channel (to prove the head is at fault).

Upon recording and playback with the reversed L and R wires, *no* intermittent recording appeared. The wires were then reversed again to their original position and the recorder tested again. Now all new recordings and playback exhibited no intermittent sound even after extended recording and playback for most of the forward direction. The conclusion was that the soldered joints at the head terminals had deteriorated in time and a poor connection developed. Since cold-soldered joints are found occasionally in all equipment, resoldering is called for whenever symptoms point to a particular unit having soldered terminals.

12-6 BUZZ IN SPEAKER

The right channel in a stereo system had an intermittent buzz that appeared only in certain musical passages or volume levels. The noise sounded almost as a rattle at times. Such symptoms are typical of loose parts in the speaker assembly and this proved to be the case. When the speakers were unplugged and connected to opposite channels, the noise came from the left channel, indicating speaker problems. (With a mono receiver a check can be made by plugging in earphones, or by attaching a test speaker.)

In this particular instance the rear panel was removed from the defective speaker and the speaker mounting inspected. A loose mounting bolt was found, and the washer rattled at a given sound level, or when a particular musical sound created a greater resonance effect in the cabinet. Evidently this bolt had not been tightened sufficiently originally because a lock washer assembly was present and these rarely come loose once tightened. A retightening of this loose bolt solved the problem.

In some instances of this type the buzzing may be caused by a loose cone at the rim of the speaker. Speaker cement is available for refastening cone sections or for repairing breaks. In rare instances a defect occurs in the voice-coil portion of the speaker. If the voice coil wires loosen, for instance, they will cause a buzz. In such an instance the speaker should be replaced if it is of an inexpensive type. If it is a high-quality expensive type, a new cone-voice coil assembly is sometimes available from the manufacturer for replacement. Also, some jobbers specialize in speaker repairs and can be found by checking advertisements in trade magazines or telephone yellow pages. If a speaker has developed a warped mounting assembly (caused by using near excessive heat, or because the frame was cracked originally or of poor quality, or dropped at some time) it is virtually impossible to undertake satisfactory repairs without incurring an expense almost equal to the price of the speaker originally. Hence, replacement is recommended.

12-7 INOPERATIVE AM-FM STEREO RADIO

A stereo radio was inoperative, though the indicator light glowed when the on-off switch was flipped. A check of the schematic showed the pilot bulb was fed from a separate winding on the power transformer. The chassis was removed and voltage tests undertaken. No circuit voltages appeared throughout the receiver except at the test point No. 1 shown in Fig. 12-3. Because no voltage appeared at No. 2, resistor R_1 was checked and found open.

A shorted filter capacitor C_2 would draw excessive currents through R_1 and cause it to burn out, but such a heavy current drain should open the series fuse in the positive-voltage line. Since the fuse tested normally when checked with an ohmmeter, it was assumed that R_1

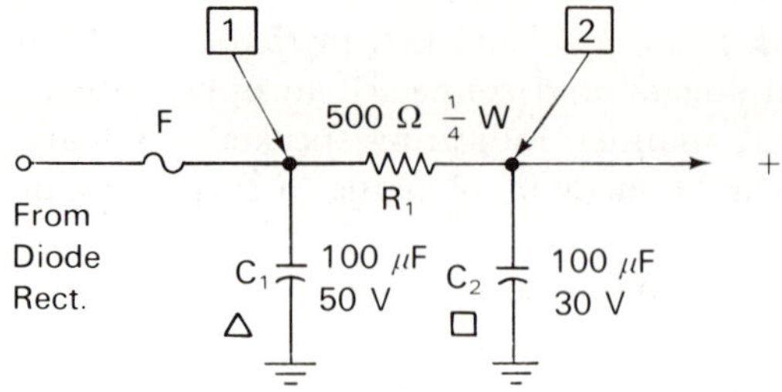

Figure 12-3 Power-supply filter section tests.

developed the open circuit during normal usage. Capacitor C_2, however, was tested with a capacitor checker and found to be in good condition. Resistor R_1 was then replaced with a ½ watt type (500 Ω) to provide for an additional safety margin. A 1 watt could also be used if there were sufficient room on the chassis.)

12-8 HUM IN RADIO SOUND

Another radio similar to that discussed in Section 12-7 had a loud hum accompanying the sound output. This is a typical symptom of a defective filter capacitor. The chassis was removed and voltages checked for a confirming clue to defective capacitors. Voltages were below normal, indicating a drain on the power-supply system (such as could occur because of a decreased resistance in a filter capacitor).

For this receiver the second filter capacitor was warm to the touch, indicating excessive current flow through it. Without additional tests, the filter capacitor was replaced (C_2 in Fig. 12-3), but with one having a 50 V rating instead of 30 V, providing for an additional safety factor. (The same 100 μF value was, of course, maintained.) With the new capacitor the hum disappeared and voltages returned to normal.

In lieu of the foregoing basic tests, of course, a capacitor checker could have been used and would have been preferable had the symptoms not been so evident. Also, as is often done, both C_1 and C_2 could have been routinely replaced to minimize the possibility of capacitor C_1 developing trouble in the near future. The practice of placing another capacitor in parallel with an existing one is not recommended. On occasion this may reduce hum by providing for additional capacitance in the filter section, but this procedures does not eliminate the possible leakage that prevails in a capacitor becoming defective. Thus, while hum may be reduced, voltages would not be at their peak and general performance would suffer. In all instances a defective capacitor should be removed and a new type installed.

For Fig. 12-3 note the identifying symbols of a triangle and a square. These are often used for flagging specific capacitors when several are within a single housing. As shown in Fig. 12-4, the bottom of a metal-can capacitor unit has output lugs also identified by the same triangle and square. Thus, C_2 is the terminal within the square identification. The wire to this terminal must be removed and a new separate single capacitor must be installed nearby. Obviously if both C_1 and C_2 are being replaced, a new unit must be obtained and the entire old unit has to be removed.

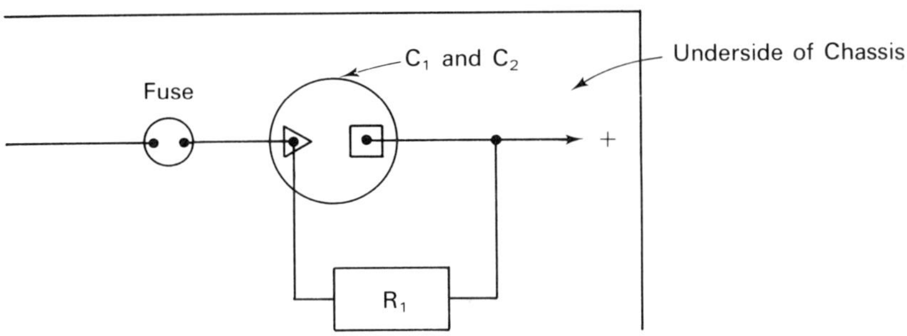

Figure 12-4 Capacitor identification factors.

12-9 WEAK FM IN PORTABLE RADIO

In a portable AM-FM radio, reception for FM stations was poor and extension of the retractable FM antenna did not appear to have much effect on performance. The back of the radio was removed, exposing connections to the FM antenna section. (The AM antenna, as in most such radios, is a separate ferrite-core loopstick.)

Inspection of the antenna assembly inside the case showed that the lead wire from the chassis connecting to the antenna rod assembly had come loose as shown in Fig. 12-5. The wire from the tuner connected to a half-ring as shown, and this metal ring fits between the case and the underside of the antenna unit as shown.

In this receiver the top lock nut had come loose by constant manipulation and closing of the antenna sections. Consequently, the metal connecting ring had worked loose and was no longer making connection. The ring was replaced and the lock nut tightened to restore full FM sensitivity.

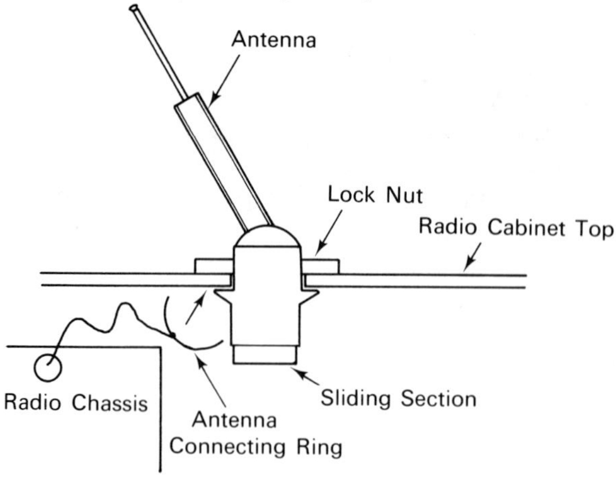

Figure 12-5 Collapsible antenna units.

12-10 WEAK AM IN AM-FM STEREO RADIO

The reception for AM was weak in a high-fidelity receiver-amplifier (AM with stereo FM). As with many of these receivers, a ferrite-core loopstick was mounted on the radio with provisions for swinging the antenna some 90° in a horizontal plane for proper orientation for best results, as shown in Fig. 12-6.

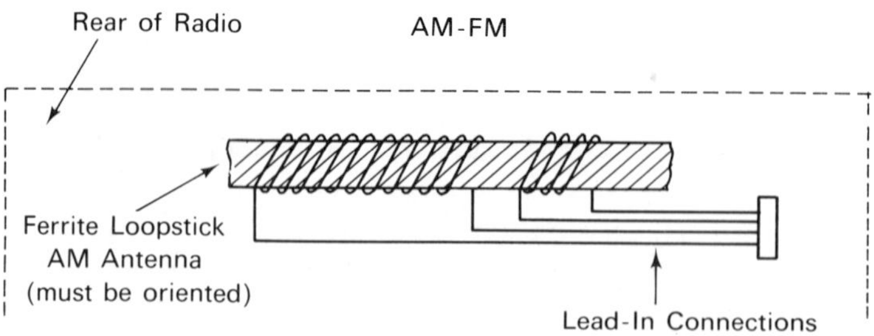

Figure 12-6 Built-in AM antenna.

Inspection of the lead-in connections from this ferrite loopstick indicated that one of the wires had broken due to constant flexing. Because this hook-up wire was very thin and the receiver had been in use some time, the wire length was disconnected from the loopstick and a new one installed. The other wires were also checked for possible weakness or breakage. With the new wire in place, full reception was restored.

Since this receiver was used in a good signal area (and in a fixed location) the loopstick is swung out to the position of best reception and usually left in place for most stations.

If reception is inadequate for a given area when the loopstick is used (or for whatever built-in FM antenna may also be used) an outdoor antenna must be installed or a coupler used with an existing TV antenna as described earlier (refer to Chapter 5 and also Fig. 5-12).

12-11 DEFECTIVE AUTO-STOP CASSETTE PLAYER

In a stereo cassette recorder-player an automatic stop feature shut off the stereo deck's power and disengaged the rollers and heads when the tape came to the end of the reel. Trouble developed and the machine would stop before the cassette tape had reached the end. As a matter of routine, the tape erase and record-play heads were cleaned. The capstan pinch roller assembly was also cleaned and the bearings oiled lightly to minimize any binding that may be contributing to the trouble. These procedures, however, were of no help, hence the chassis was removed to gain access to the automatic-stop mechanism.

For this cassette deck the end-of-tape sensing consisted of the units shown in Fig. 12-7. As long as the drive shafts of the tapes were in motion, a linking belt connected to a perforated disk shaft caused it to rotate. A small light bulb was positioned to direct illumination through the perforations to a photo-diode sensing unit. Thus, while the disk rotated, the moving perforations permitted light to come through in periodic segments, producing a form of alternating current. When the tape reached the end, the disk would stop rotating and the lack of ac arriving at the sensing diode caused a relay to trip and shut off the power and disengage the heads.

Inspection of this unit showed a binding in the bearings of the disk assembly, causing belt slippage on occasion. (A loose belt could also

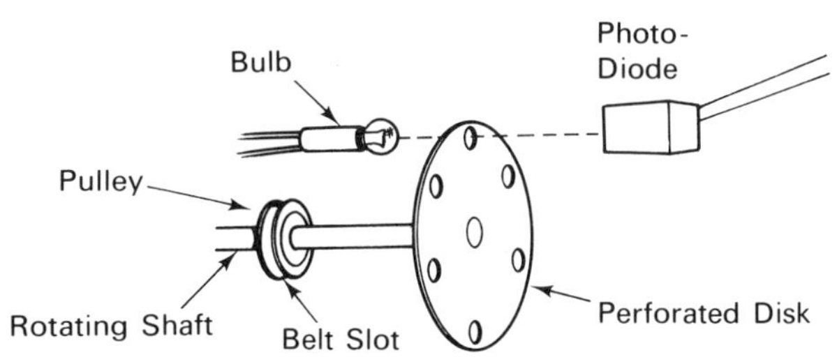

Figure 12-7 Tape-end sensing units.

cause shut off because sufficient pulley pressure could not be maintained. If the belt is loose, a new one must be obtained from the manufacturer or jobber, since a stretched one cannot be repaired.) Sometimes oil may get on the belts and cause slippage, in which case a cleaning fluid must be used to remove all oil or grease from both the pulley and the entire length of the belt.

For the deck in question, however, proper operation was restored by a drop of oil in the bearings associated with the tape-end sensing units. Also, the bulb had a coating of dust on it and both this unit and the photo diode were cleaned to increase sensitivity.

12-12 WEAK AND FUZZY LEFT CHANNEL IN PHONO

In a stereo record player the arm was accidentally dropped on the turntable and the cartridge needle assembly damaged. The owner installed a new cartridge, but the left channel sound was weak and fuzzy. Inspection showed the new cartridge did not fit as well as the original one and the needle was not at perfect right angles to the record, causing an off-balance condition between the two channels (favoring one channel while picking up a distorted and weak signal for the other).

The tilt of the cartridge was corrected by adding a thin washer between cartridge and shell where the bolt fastened the assembly on one side. A carpenter's level was also used as shown in Fig. 12-8 to ascertain that the turntable was level both along the front and down the sides to minimize groove skipping and to reduce the groove pressure against one wall of the record because of turntable tilt. These procedures both corrected the trouble and restored equal reception for both channels.

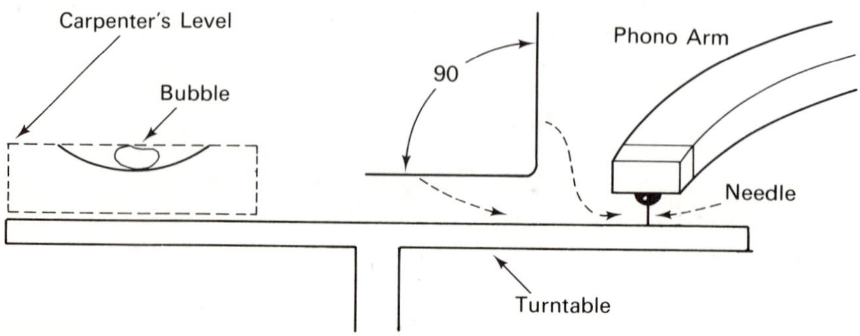

Figure 12-8 Turntable correction factors.

12-13 RADIO NOISE INTERFERENCE

A high noise level interference was present for AM only in an AM-FM receiver. The loopstick ferrite-rod antenna was disconnected to ascertain whether the noise was picked up by it, was coming in the ac line, or was caused within the set. Since the noise was still present with the antenna leads open, the leads were reconnected. The radio was brought to the repair shop where it performed satisfactorily without any noise interference. A commercial line filter was installed and this was successful in eliminating the interference.

If commercial line filters are not readily available, paper, mica or ceramic capacitors can be placed across the line input as shown in Fig. 12-9. Such capacitors must be of the nonpolarity type and not the polarized electrolytics used on dc. The arrangement in Fig. 12-9A consists of two capacitors, with the center common connection grounded to the radio chassis. A single capacitor can be used as shown in Fig. 12-9B when it is known that the ac line has a good ground for one lead. Values can range from 0.01 μF to 0.25 μF, with the larger values providing for a greater shunt for interference signals. Working-voltage ratings should be above 200 V.

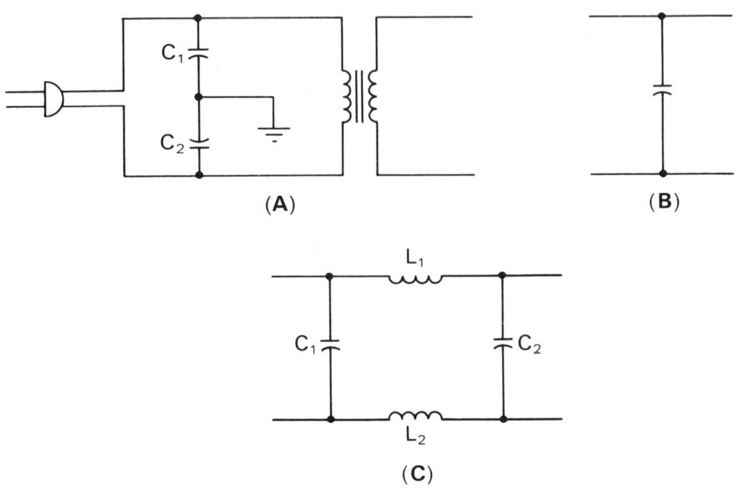

Figure 12-9 Noise filtering in ac line.

Commercial types generally use series inductors as well as shunt capacitors as shown in Fig. 12-9C for more effective noise filtering. Higher-frequency signals find an increasing reactance (opposition to entry) in the inductors, while at the same time being subjected to decreasing shunt reactances in the capacitors. Inductors, however, must have wire sizes sufficiently large in diameter to carry full currents for the power handled, otherwise they will overheat and may burn out.

12-14 EXCESSIVE RECORD WEAR

After records were played a few times on a phono, the record showed signs of wear and developed a high background scratch noise. For this phonograph, the stylus pressure measured 3 grams, and though a lower pressure would be desirable, the 3 grams should not cause the excessive record wear that was experienced. The cartridge was removed and the needle point viewed through a 50-power hand-held magnifier (available from jobbers), as shown in Fig. 12-10A. This inspection disclosed a chipped needle point (synthetic sapphire). This sometimes occurs if the tone arm is dropped on a hard-surfaced turntable or hard top.

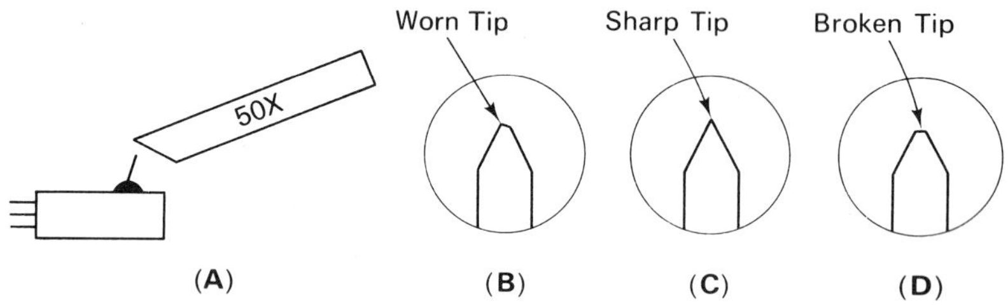

Figure 12-10 Phone needle checks.

A replacement needle restored proper operation. As shown in Fig. 12-10B, a worn needle tip has a flat section on one side of the converging sides, as contrasted to the point shown in Fig. 12-10C. A broken or chipped needle point produces a ragged or uneven tip that grinds away the record walls.

12-15 LOSS OF STEREO IN FM

In a stereo receiver the sound output was mono only for all stations and the stereo indicator light did not function. Inspection of the stereo-mono switch did not disclose any faults, hence the trouble was assumed to be in circuit components associated with the 19 kHz and 38 kHz amplifiers, since troubles in these circuits would not necessarily prevent the mono signals from arriving at the bridge detector (see Figs. 9-2, 9-3 and 9-4).

A scope check indicated the arrival of signals at the output of the 19 kHz amplifier transformer (Fig. 12-11). The diodes and resistors were checked with an ohmmeter and found to be normal. No output was observed from the 38 kHz amplifier transistor. Because components checked out all right, the transistor was suspected. Upon removal and testing it was found to have opened on the emitter-collector side. (This would also cause inoperation of the stereo indicator light shown in Fig. 9-3 and associated with transistor Q_3.) After replacement of this transistor, normal reception in stereo was restored.

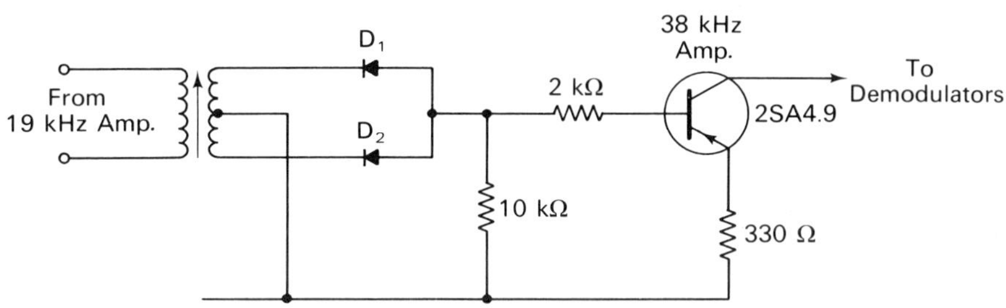

Figure 12-11 Doubler diodes and amplifier components.

12-16 INTERMITTENT AND NOISY STEREO SOUND

In a stereo amplifier the sound output was noisy and intermittent. When the volume control was adjusted there was a considerable amount of noise from the speakers, indicating a faulty volume control. This volume control was a dual type of 30 kΩ each (so a single shaft and knob can be used to control both channels simultaneously). Sometimes such controls become noisy because of dust or dirt entering the case, even though the resistance strip may be intact. Hence, use of spray-can noise-reducing chemicals is often effective in eliminating such troubles, temporarily at least. (Various types of cleaner chemicals for specific purposes are available from jobbers.)

Sometimes running the cleaner fluid down the shaft (without chassis removal) suffices. For the amplifier in question, however, this was insufficient, so the chassis was removed to permit access to the volume-control sections. Cleaner fluid was run down the slots where the leads entered the controls. The control knob was turned while the fluid was run in and the noise stopped. From the initial symptoms, however, it appeared that this was a temporary measure and that the volume controls should be replaced. Since these were dual controls (with specific taper and with specific dimensions) a factory replacement was then ordered.

Similar procedures can be undertaken for noisy tone controls, bass boost controls and so forth. For cleaning band-switching sections, TV and FM tuners and other sections where high-frequency signals are involved, make certain the cleaner chemicals do not introduce losses by decreasing circuit resistances to VHF and UHF.

12-17 POOR RECEPTION IN FM RADIO

In an FM portable receiver signal pick-up was poor, even though it was used in an area that had a number of local FM stations. Results were the same whether the radio was used with internal batteries or on the ac line. (With such receivers, of course, weak reception is usually the fault of aged batteries with low voltages.) When the volume control was turned up on this receiver, the increased level of background noise gave some evidence that the fault was not in the audio amplifiers, but rather in the circuitry prior to the FM detector.

The chassis was removed and voltages checked throughout the receiver and compared with the voltages specified on the schematic for this set. All voltages were normal, which would also tend to indicate that associate components were all right (shorted or open resistors as well as shorted capacitors would upset voltage readings). Transistors were also tested and found to be satisfactory. Finally signal tracing was undertaken by injecting a signal into each stage and observing the peak-to-peak output on a scope (see Fig. 6-2). When the signal generator was adjusted for the 10.7 MHz IF signal, the output observed on the scope was lower than when the signal generator output was changed in frequency below the required 10.7 MHz.

Thus, misalignment of the IF stages was indicated. Because the receiver had operated satisfactorily for over a year and all components and transistors checked out satisfactorily, there would be no reason for a sudden shift of IF frequency alignment unless someone had tampered with the IF transformer adjustable-core settings (see Chapter 7, Sections 7-1 and 7-5). (Too often the inexperienced in electronic servicing will attempt adjustment of the IF transformer cores without use of alignment equipment in a mistaken effort to "improve" the reception.) For the particular receiver in question, the tuner was found to track properly, and a realignment of the IF stages then restored the receiver to peak performance.

12-18 STEREO FM SOUND LOSS (CASE 1)

In a stereo FM receiver there was no sound output, though faint speaker noises indicated that the receiver was not completely dead. The tuning indicator meter did not function, indicating that the trouble would be in the stages preceding the third IF, which feeds the tuning indicator as illustrated in Fig. 12-12. (Shorts in the power supply, of course, could cause loss of reception because such shorts would affect all stages.)

The chassis was removed and initial voltages were read for the IF stages. At the collector of the second IF amplifier, however, no voltage appeared, indicating either an open primary IF transformer winding or an open 270 Ω resistor. A continuity check showed an open series resistor. As a matter of precaution the second IF transistor was also checked, because a short between emitter and collector would cause excessive currents to flow and would overheat and possibly burn out the series 270-Ω resistor. The transistor checker indicated the suspected short between emitter and collector. Thus, replacement of the transistor and the series resistor restored operation. (The series resistor was a ¼-W type and this rating was not increased for the new resistor because the lower rating helped protect the power supply from shorts by opening up for excessive currents.)

12-19 STEREO FM SOUND LOSS (CASE 2)

A stereo FM receiver had no sound output, though background noise increased when the volume control was turned up, indicating the set was not totally inoperative. When the station selector knob was tuned, the tuning indicator meter needle swung up and down the scale as various stations were tuned in, indicating the receiver was passing the signals up to and including the third IF amplifier for this receiver (see Fig. 12-12).

For this receiver an IC module was used that combined the ratio FM detector and the first audio amplifier as shown in Fig. 8-15. A scope test showed signal input to the IC module, but no sound output. Resistors and capacitors associated with this module were checked to see whether faulty components associated with the IC module were the cause of the trouble. These checked out satisfactorily, and proper voltages were available from the power supply, indicating a faulty IC module. An identical type was obtained and installed and normal reception was again obtained for this receiver.

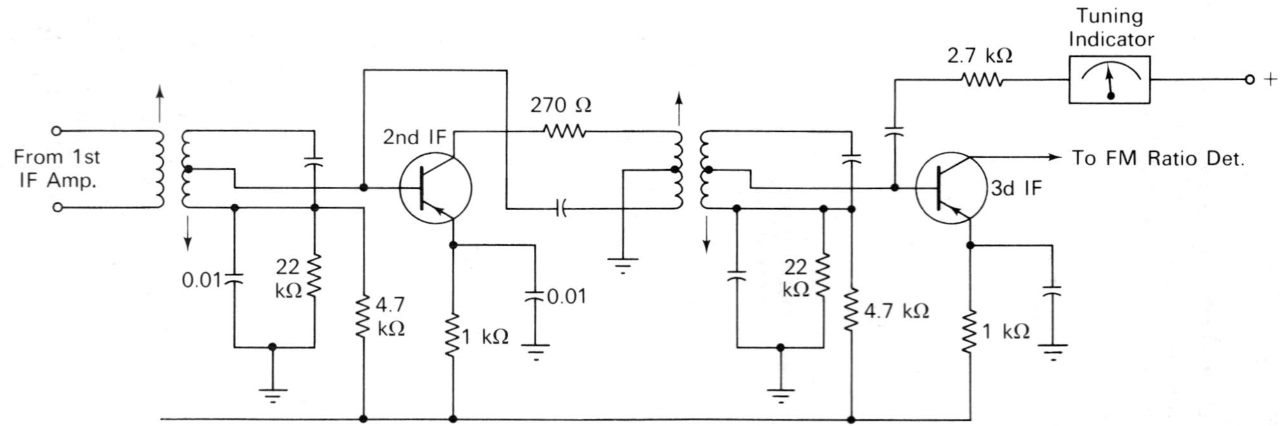

Figure 12-12 Stereo FM IF stages.

12-20 STEREO FM SOUND LOSS (CASE 3)

A portable stereo radio with two built-in speakers was dropped on a concrete sidewalk after which the left channel sound was very weak and the right channel was completely inoperative. The fall had broken the antenna connections and damaged the right-hand speaker grille. Plastic cement was used to repair the cabinet, and connections were resoldered to the antenna (see Fig. 12-5).

Speaker connections were inspected and it was found that the right-channel speaker wires had been torn off. After the wires were replaced and the set tested again, sound output from the right channel was fuzzy and distorted, though the left-channel sound was now satisfactory. Stereo earphones were plugged in and the sound output from each channel was clear and had good volume. The stereo earphone jack was inspected (since it shut off the speakers when the stereo earphone plug was inserted) and connections were found to be good. It was suspected that the right speaker frame had been bent slightly and the voice coil was rubbing against the pole pieces of the speaker magnet. A replacement speaker was obtained and installed, and normal reception was restored.

This case indicates the ever-present possibility of multiple troubles existing simultaneously in any electronic device, even though not subjected to physical shock. It is possible, even though remotely so, that an audio transistor could become defective at the same time that a capacitor or resistor opens in the tuner. Also, another rare occurrence is that a newly acquired replacement component may be defective. Many technicians insist on checking all new resistors, capacitors and transistors before installation.

12-21 STEREO FM SOUND LOSS (CASE 4)

During recording with a cassette stereo tape deck, both VU meters were activated, showing that the signals were getting to the recording heads. On playback, however, only the right channel produced a sound output. The output coaxial cables were unplugged and reversed, but the left channel still produced no sound. A commercial test tape was played and still there was no sound from the left channel, indicating the trouble was probably caused by defects in the playback circuitry and that the deck was recording properly.

The chassis was removed and an oscilloscope was used to check for signals while the test tape was played. Scope patterns were displayed when the scope probes were placed at the collectors of each output power transistor. The wiring leading from the output transistors to the output coaxial socket was inspected carefully, and appeared to be correct. One wired connection to the soldering lug of the output socket appeared corroded. When this wire was flexed, it appeared loose within the soldering joint. The joint was resoldered and normal output was again obtained from this socket, indicating a cold-soldered joint was present that had corroded sufficiently in time to prevent electrical contact.

12-22 STEREO FM SOUND LOSS (CASE 5)

In a stereo FM receiver there was no sound output from the left channel for FM music, tape input or phono input. Speaker plugs were reversed to check whether speaker faults were present, but the right-channel speaker also did not produce sound when plugged into the left-channel output. A check with stereo earphones also indicated a dead left channel.

The chassis was removed and voltage checks made at various test points through the receiver, with no off-values found. In the audio sections for each channel, voltages were identical at similar points. Next, a signal was injected into the tuner and a scope used at the output of the ratio detector. Here a signal appeared and was traced to the dual-potentiometer volume control. A 1 μF, 30 V capacitor coupled the signal energy from the volume control to the base input of each audio transistor, as shown for one of the channels in Fig. 12-13. When the scope probes were placed between the base of the transistor and ground for each

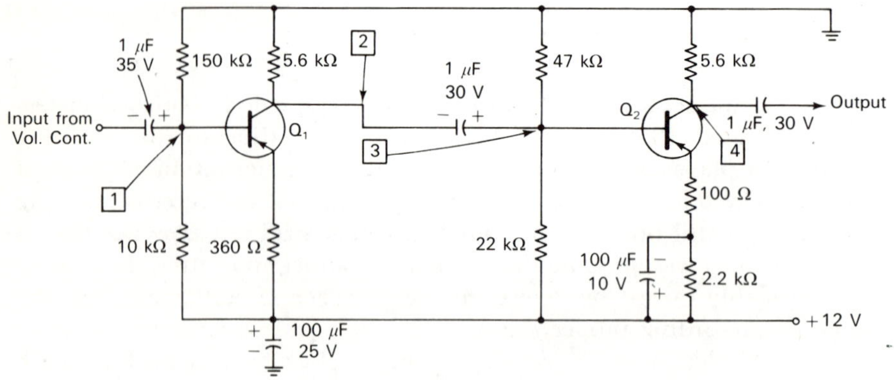

Figure 12-13 Test points for fault location.

channel, the signal again appeared, indicating no faults at test point No. 1 for each channel.

A scope test was made at test point No. 2 for Q_1 (left channel) and Q_{101} (right channel). Again the sound signal appeared at each place. However, when the scope probes were placed at test point No. 3 and ground, no signal appeared for the left channel circuit, but was present at the right-channel input to the second resistor (Q_2 for the left channel and Q_{102} for the right). Thus the 1 μF coupling capacitor was suspected of being open, and when tested this was found to be true. When a new 1 μF, 30 V capacitor was installed, normal sound was restored to the left channel.

Care must be taken in replacing capacitors of this type because in many instances the larger-value types are electrolytics and hence polarities must be observed to avoid malfunction and burnout. In most schematics the polarities are marked, but if not identified on the schematic, the old capacitor should be inspected to note how it was wired into the circuit. (For *npn* transistors, instead of the *pnp* transistors shown in Fig. 12-13, the polarities of the capacitors would be the reverse of those that are shown.)

If the defective capacitor had a voltage rating close to the actual circuit voltages present it is advisable to replace it with a higher voltage rating type, though of the same capacitance value.

12-23 POOR B/W PICTURE QUALITY

A 19-inch black-and-white receiver (tube type) had a blurred picture with washed-out contrast and poor brilliancy. Picture highlights had a silvery appearance, indicative of declining emission in the picture tube. Control adjustments were tried, but did not improve the picture. The rear panel of the receiver was removed and a separate TV-ac line attached so the receiver could be turned on. A high-voltage check was routinely made and the voltage was normal. Next, a tube checker was applied to the picture tube and, as suspected, emission was low. A picture-tube booster (brightener) was tried and this increased emission sufficiently to restore good brilliance and sharpness.

A typical picture booster of this type is shown in Fig. 12-14A and consists of a picture-tube socket and plug, plus a housing containing an autotransformer (as shown in Fig. 12-14B). This transformer steps up the filament voltage for the picture tube to raise the temperature, and thus increase the cathode emission. For a 6.3 V filament the step-up voltage may be as high as 8 V. Installation is simple: the picture-tube socket is

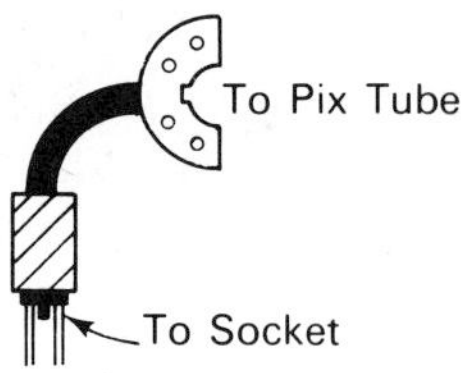

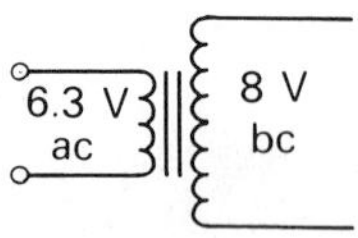

Figure 12-14 Picture-tube booster.

removed and the booster socket installed. The booster plug is inserted into the socket that was removed from the tube.

When installing such a booster, make sure it is of a type suitable to the particular picture-tube in use. Several models are available, some for the series-string tube system and others for parallel. For all solid-state receivers only the tube filament potential is a factor. The picture tube quality can be increased to last several more years by such units.

12-24 COLOR PICTURE CONTAMINATION

In a 25-inch solid-state color receiver color blotches appeared whenever a single-color background was being viewed. It was evident that the receiver was at fault because the false color spots always appeared at the same place on the screen. Such a condition was usually caused by incorrect purity-ring adjustments (see Fig. 10-16). Accordingly, the purity rings were adjusted as described in Sec. 10-10. This procedure helped, but some contamination still remained at the picture-tube sides. Because this could be caused by a magnetized area near the picture tube, a degaussing coil was used (available from jobbers) to demagnetize the areas all around the tube.

As shown in Fig. 12-15, the degausser usually consists of a coil ring encased in plastic, with a push-button switch and a length of cord for plugging into the ac line. The button is pushed and held in to generate the magnetic field. The coil is then moved all over the tube face, sides and cabinet top (with or without the set turned on). After completely sweeping the entire tube-face area, the degaussing coil is removed some distance away (about six feet) *while keeping the coil energized.* The push-button switch is not released until the coil is this distance, to avoid remagnetizing picture tube sectors by the collapsing fields of the magnetism when the button is released.

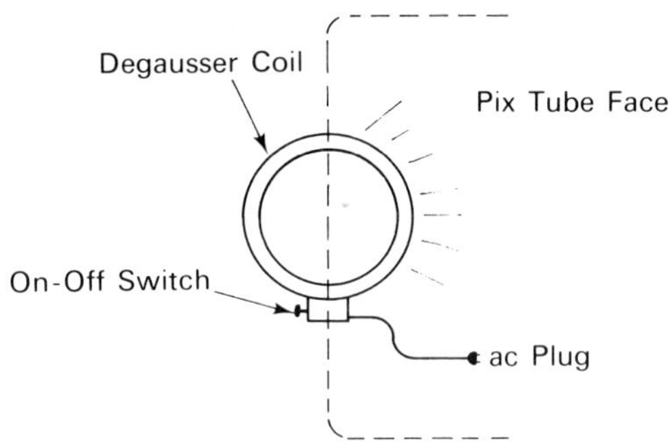

Figure 12-15 Color tube demagnetizing.

While sets have built-in degaussing coils surrounding the picture tube, on occasion metallic supports will pick up stray magnetism (from vacuum cleaners, etc.) and require degaussing. (The procedure should be routine for every service call.)

12-25 POOR FOCUS

In a portable color receiver the focus was poor and adjustment of the focus control failed to obtain the horizontal-line trace that should be visible on the screen when focus is correct. A high-voltage probe was used in conjunction with a VOM and the voltage measured at the focus electrode of the picture tube. The voltage read only 3 kV instead of the 5 kV needed for this particular receiver (see Fig. 10-13).

For this receiver stacked diodes (series) were used in a plug-in assembly. Installation of a new unit corrected the poor focus condition and restored the focus voltage to that required.

12-26 PICTURE BLOOMING IN B/W SET

In a black-and-white receiver (vacuum-tube type) the entire picture was enlarged abnormally and could not be reduced to proper size with any control adjustment. This condition is referred to as *blooming* and is usually caused by a *decrease* in high voltage. When the high voltage is below normal the sweep amplifiers of the tube can sweep the beam to a greater extent because beam velocity is less with inadequate high-voltage on the picture-tube's second anode (inner coating).

The horizontal output tube and high-voltage tube for this receiver were found to be low on emission and replaced. The new tubes increased the high voltage, picture size was reduced to normal and full brightness was restored. As shown in Fig. 12-16, a damper tube is also associated with the output circuitry and this tube should be checked when high-voltage troubles occur. The tube also contributes to the voltage boost applied to the voltage obtained from the low-voltage supply. Thus, if the low voltage declines below that specified for a particular receiver, the fault could be caused by a low-emission damper tube.

Note that for this receiver a fuse is in series with the screen-grid of the horizontal output amplifier. This fuse will open up if the screen-grid bypass capacitor shorts, thus protecting the power supply system. Overloads and defects in the horizontal output tube would also cause the fuse to blow. Hence, if an open fuse is found, circuit components should be checked.

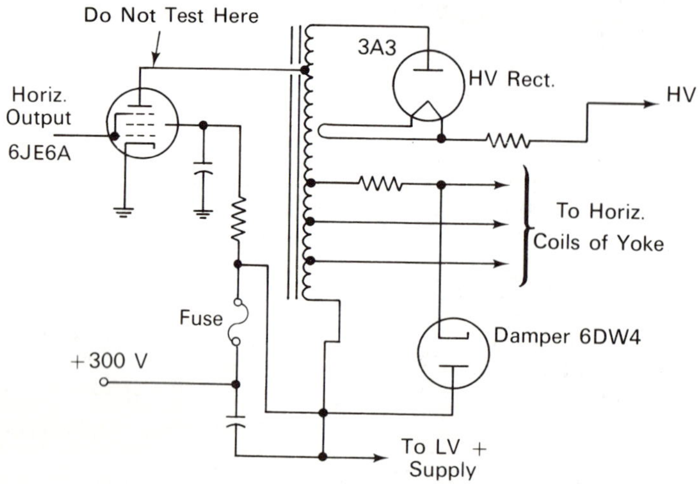

Figure 12-16 Tube-type horizontal and HV circuitry.

12-27 POOR COLOR RECEPTION

In a color receiver good color reception could not be obtained with control adjustments. In color scenes red fringing would appear beside faces or small objects on the screen. Pure white was never obtained for any portions of the picture (white shirts, etc.). Instead, such areas were off-white and gray. Since these symptoms indicated poor convergence, a dot-bar generator was used and it indicated that convergence adjustments were off by a considerable extent.

Poor convergence could be caused by circuitry faults, but in this instance the convergence controls functioned normally and convergence procedures restored good picture quality without fringing, and with excellent white areas when such were telecast. Thus, this receiver had either been tampered with or was not properly converged initially.

As part of routine color servicing, purity was checked and degaussing procedures applied to the tube face area. Proper levels of blue, green and red drive and screen controls were also checked and adjustments made to the controls where necessary.

12-28 CHASSIS BUZZ

In a solid-state color receiver there was a noticeable buzz at all times, even when the volume control was turned completely down. Because the buzz was still present when sound was turned down, the buzz was obviously caused by some component on the chassis. The back was removed and it was immediately apparent the buzz was produced by a small transformer. A check of the schematic showed this to be the filament transformer for the picture tube.

The transformer was unbolted from the chassis and inspected. The buzz was caused by loose laminations which vibrated when ac flowed in the windings. The laminations were supposed to be held tightly by the metal-case housing and evidentally these were not clamped tightly enough. Pliers were used to press the metal housing portions that held the laminations. This resulted in a tighter hold and eliminated the buzz. If this had failed, a new transformer would be required because it is often difficult to make repairs without considerable trouble. (Speaker cement could be tried on the laminations. After drying and hardening, the cement would hold the laminations tightly and prevent the buzz. However, if the transformer runs fairly warm, some trouble may be experienced with many types of cement. Thus, it is better to obtain a new transformer if troubles persist.)

12-29 SOUND LOSS IN TV

In a television receiver there was no sound output for any channel. The volume control was ineffective in producing sound output. There was a good picture present on the screen, however, and this indicated that both the low- and high-voltage power supplies were functioning correctly.

The back was removed from the receiver and a line cord attached to the input ac receptacle. One side of the speaker was disconnected and a test speaker applied to the audio output terminals indicated in Fig.

12-17. The substitute speaker failed to produce sound output, so the original speaker was reconnected and a scope used to signal-trace the audio. For this receiver two IC modules were used in the sound section. One module contained the IF stages and FM detector, followed by the volume control and the second IC module shown in Fig. 12-17. Sound signals appeared at the input to the audio module shown (sliding volume-control arm) but no sound was available from the output. A voltage reading was made at input lug No. 2 of the module and ground. According to the schematic, this was supposed to read +25 V, but only approximately 16 V appeared here. When the module was removed from the chassis, the voltage rose to slightly over 25 V, indicating a leakage and fault within the module. (External capacitors checked out all right.) Accordingly, a new replacement module was obtained and installed and full sound reception was regained.

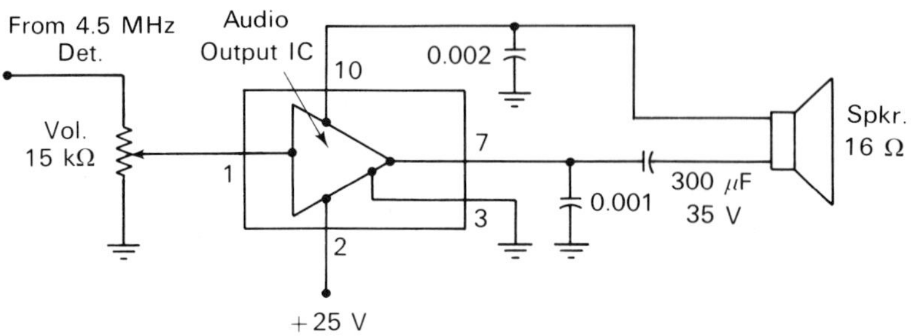

Figure 12-17 Audio module external components.

12-30 OCCASIONAL SYNC LOSS

In a television receiver the picture would tear on occasion and roll vertically. After a few seconds the picture would lock in again and remain so for five or ten minutes. When such intermittent conditions are present it is usually difficult to pinpoint the trouble, particularly if the fault only appears momentarily every hour or so. However, with this receiver, the sync loss (both vertically and horizontally) happened often enough that meters could be applied to the circuit for observation of changes that occur when the fault appears.

Before test instruments were used, however, routine checks were made to see that the vertical and horizontal hold controls were adjusted for best lock conditions. These were found to be satisfactory. The back was removed from the receiver and voltage tests were made at the sync separator circuitry (because both vertical and horizontal sync were affected).

All dc voltages were normal, which is to be expected during normal operation. The schematic for this receiver gave the waveforms that were to be expected at the sync separator (plus the peak-to-peak voltages) as shown in Fig. 12-18 (see also Figs. 6-8 and 6-9). A scope check at the base and collector also showed normal waveforms and PP voltages. The oscilloscope was left at the collection (as against ground) so that the waveform could be observed when the intermittent condition occurred. When sync was lost the waveform distorted as shown in Fig. 12-19, and the PP voltage dropped to approximately 45 V.

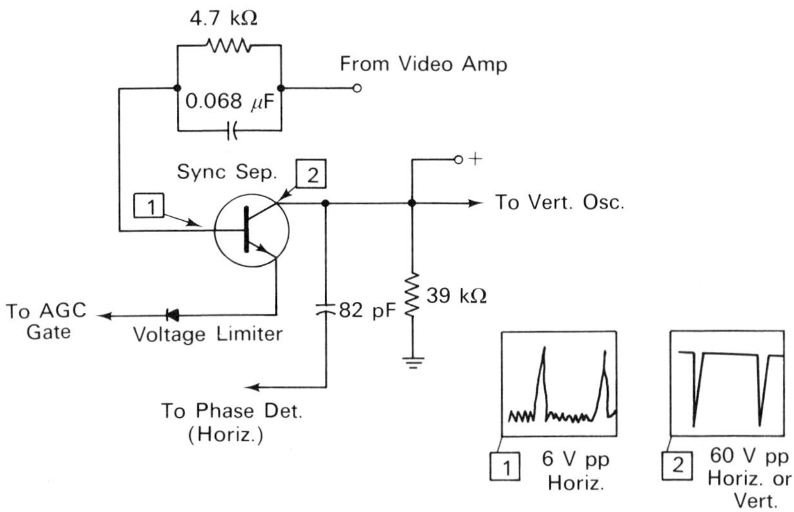

Figure 12-18 Sync separator waveforms.

As an additional check, the scope was connected between the base input and ground and observed during the sync-loss condition. Here, however, the waveform amplitude changed only slightly during the trouble, indicating the fault was probably in the transistor or associated components at the collector output. (Some shunting effect at the collector side would be felt to some degree at the input, particularly if the transistor characteristics were not what they should be for proper operation.) Next the transistor was tested with a solid-state checker, but was normal. Thus, the transistor was also normal until the intermittent fault accurred. A new transistor was installed and the set no longer exhibited the intermittent conditions.

In intermittent circuitry, voltmeters as well as scopes can be placed across certain test points and left in place for checking conditions when the intermittent problem occurs. If the trouble cannot be found

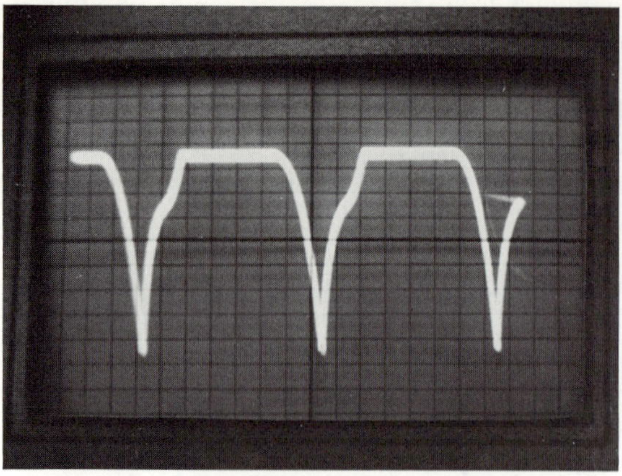

Figure 12-19 Negative pulse distortion.

readily, all components that are slightly off should be replaced in circuitry associated with the fault. (If the sound is intermittent, for instance, check audio components. If only vertical sync is affected, test the vertical oscillator and vertical amplifier circuitry, etc.) (See also Fig. 4-22 and 10-8.)

Some technicians use a variable line voltage transformer to raise the input voltage slightly and thus overload components to a small degree. This sometimes hastens the intermittent condition or causes a burnout of the defective component for quick localization. Others have placed cartons over the receiver chassis to increase heat on components to hasten the process. These procedures, however, may also damage other parts and are not recommended.

12-31 POOR COLOR FLESH TONES

In a television color receiver using a single-gun picture tube, the picture had sufficient contrast and brightness but color tones were not true. Adjustments of the drive and background controls were made but did not help, because the red background control did not operate properly and could not produce sufficient red on the screen. As shown in Fig. 12-20, these controls at the input of the picture tube correspond to similar ones in the three-gun television where screen controls are used for similar functions (see Fig. 10-15 and Sec. 10-10).

Resistor R_5 was tested and found to be defective. Replacement was made and both the drive and background controls were readjusted for uniform beam velocity for the three colors. As a matter of routine, focus and convergence voltages were checked and found to be normal. Troubles

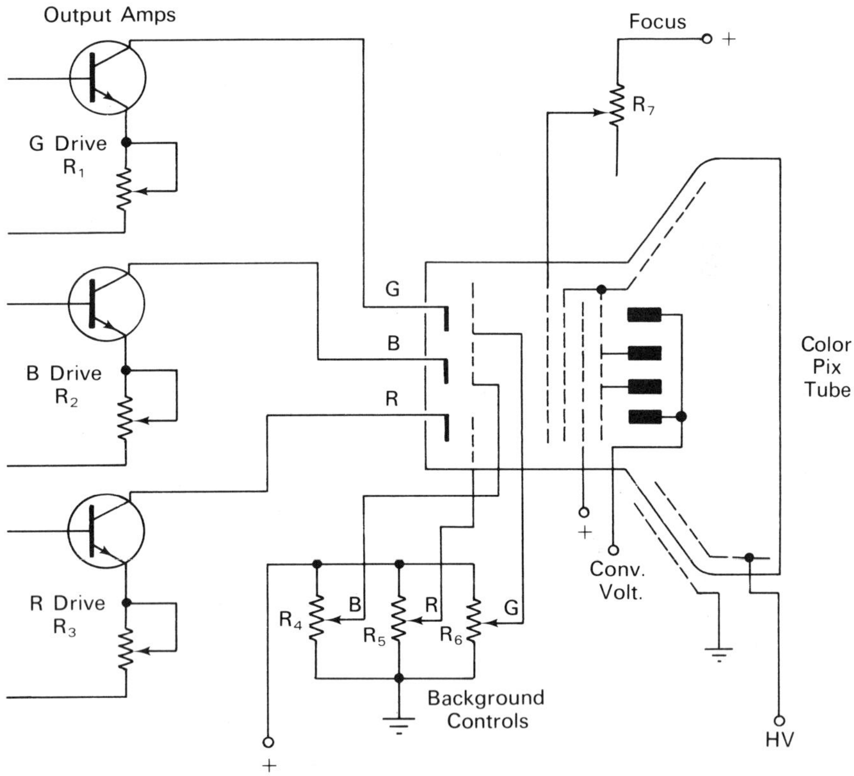

Figure 12-20 Single-gun color tube input circuitry.

of this type could also be caused by faulty resistors in the drive circuits (R_1 through R_3) as well as defective output amplifier transistors, and all these must be tested if the background control resistors had been found to be correctly adjusted and operating properly.

12-32 AUTOMATIC COLOR CONTROL FAILURE

In a color receiver with automatic color control (ACC) the colors did not hold true when the automatic control button was depressed. As shown in Fig. 12-21 this receiver uses a diode and transistor amplifier to form the ACC circuit. The ACC functions similarly to AGC, but uses the 3.58 MHz signal instead of sync pulses. Diode D_1 rectifies the signal and applies it across R_1 and hence between base and ground. The degree of conduction of the transistor then regulates bias to the bandpass amplifier and maintains a fairly constant color level.

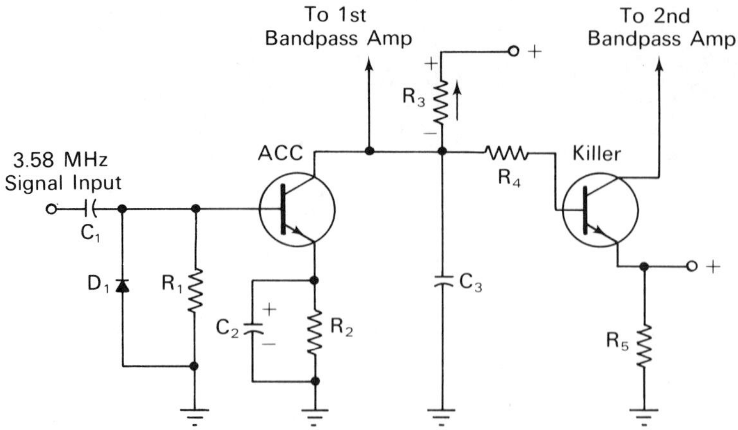

Figure 12-21 Automatic color control and color killer.

The diode, resistors and capacitors associated with this circuit were tested and found to be satisfactory. Next, the transistor was checked and found to be below normal in characteristics. A replacement transistor restored proper ACC function.

12-33 REPEAT LINES AND POOR LINEARITY

In a black-and-white receiver there were repeat lines to the right of objects and all scenes were elongated vertically. These conditions are evident in the picture shown in Fig. 12-22. The vertical elongation was

Figure 12-22 Repeat lines in picture.

caused by a misadjusted vertical height control. The vertical height control was adjusted to bring the vertical scan within the picture-tube frame. This adjustment, however, threw off the linearity slightly and required touching up the linearity and readjusting the height control.

The repeat lines were caused by an improperly oriented antenna that was not always pointed toward the station to be received. A new high-gain antenna and an antenna rotator were installed and satisfactory reception was obtained (see also Figures 5-7 and 5-11).

12-34 DARK HORIZONTAL BAR ON SCREEN

In a hybrid black-and-white television receiver (using some solid-state devices and some vacuum tubes) a dark horizontal bar appeared on the screen as shown in Fig. 12-23. This is a typical symptom of 60 Hz hum entering the video amplifier stages. (With 120 Hz two bars would be visible.) Accordingly, the video amplifier tubes were checked and it was found that the first video amplifier tube had developed a cathode-filament short. A new tube restored good reception.

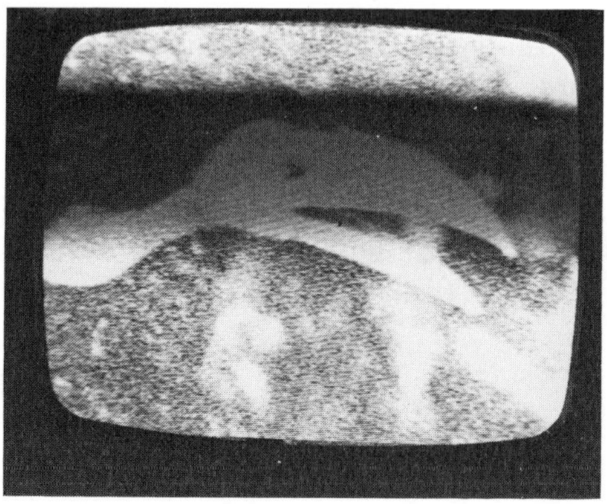

Figure 12-23 Hum bar in picture.

12-35 SOUND INTERFERENCE

In a black-and-white receiver fine lines and segments appeared on the screen and varied in intensity and position with the sound from the receiver. The fine tuning control was readjusted, but the trouble persisted on each station tuned in. The back was removed and a substitute line cord attached so the receiver could be operated.

The 41.25 MHz (own sound) trap was adjusted until the sound interference was eliminated (see Fig. 7-6). (If trap adjustments do not help, circuitry associated with the traps should be checked. Also, tuner tracking and alignment may need slight retouching for proper bandpass characteristics to reject sound interference and adjacent-channel interference.)

12-36 NO PICTURE OR SOUND

In a color receiver there was no picture or sound when the receiver switch was turned to the on position. Since the pilot light was also out, it appeared no ac power was being applied to the receiver. A portable troubleshooting lamp was plugged into the wall socket used for the television. Because the light operated satisfactorily it eliminated the wall socket as a possible cause of the inoperative TV. Next the ac cord to the receiver was inspected and appeared to be intact. An ohmmeter was applied to the prongs of the ac cord and the television on-off switch flipped. No continuity was obtained, indicating an open circuit in the ac input to the set.

The back was removed from the set and a substitute cord attached to the ac input, but the receiver remained inoperative. (The fuse in the receiver was not blown.) The ac cord was again plugged into the wall socket and an ac voltmeter used to check the on-off switch. This switch was part of the volume control and the voltmeter read ac to the input of the switch (as against the other ac line not attached to the switch). No voltage was read at the output of the switch, however. Hence, a new volume-control-switch assembly was obtained and installed, which corrected the trouble.

12-37 INTERMITTENT BRIGHTNESS

In a color receiver the brightness varies slightly for every station. Adjustment of the brightness control to different levels of brightness did not stop the fluctuating brightness. The brightness would stabilize for a few minutes then vary again.

In this receiver the contrast, brightness, tint and color controls connected to a wiring harness as shown in Fig. 12-24, which in turn connected to a 12-prong plug that was inserted into the appropriate socket on the chassis. When this cable was moved slightly it initiated the intermittent condition. When the plug end was held rigidly and the cable flexed, however, it had no effect on the intermittent condition. Hence, it appeared

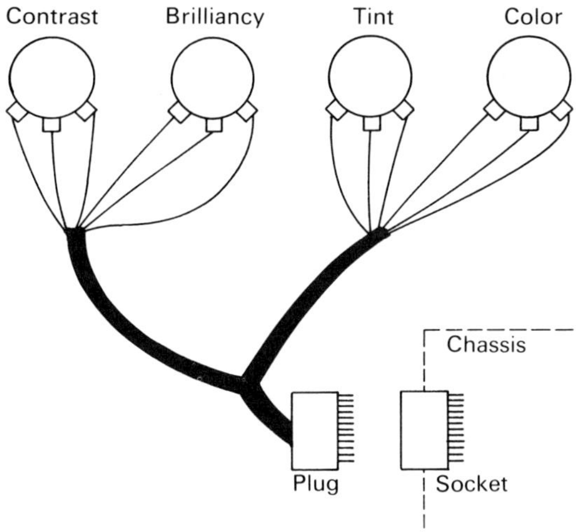

Figure 12-24 Control-cable factors.

the plug and socket assembly were not making good contact for the brilliancy control section.

The plug was removed from the socket and the prongs cleaned with contact cleaning fluid and reinserted firmly. This eliminated the intermittent brilliancy condition.

In another receiver a similar symptom of intermittent brilliancy was caused by a defective resistance strip in the brilliancy-control potentiometer. A spray cleaner of the type used for volume controls remedied the condition, but because of the voltages involved this was considered a temporary measure until an exact replacement part could be obtained.

12-38 POOR UHF RECEPTION

The purchaser of a new 19-inch portable color receiver complained of poor UHF reception. The receiver had performed well at the store, however, hence the owner's location was checked out. A new antenna had been installed and it had both VHF and UHF elements. At the receiver, however, two conditions were found that caused poor reception.

One factor was that the incoming twin lead had not been cut off, but about 12 feet of the cable was curled up in back of the receiver. Such an excessive amount of coiled lead-in can cause a significant loss of UHF signals. In addition, the set owner did not have a VHF-UHF splitter and had shunted the two inputs together at the receiver, thus shunting each input circuit by the other and decreasing input impedance. A VHF-UHF splitter was installed (see Fig. 5-12) and the excessive amount of lead-in cable was cut down so that just enough slack remained for repositioning the set slightly if needed. This resulted in a considerable improvement in both VHF and UHF reception, though optimum results were not achieved because the antenna was in a fixed position. It was recommended that an antenna rotor be installed if additional improvements were required.

12-39 EXCESSIVE CONTRAST

In a console receiver the contrast was excessively dark for most local stations and the contrast control or AGC control was ineffective in correcting the condition. Such excessive contrast may be caused by faults in the AGC system in which the AGC no longer prevents the overloading that could occur for strong local stations.

For this receiver an AGC gate transistor and an AGC amplifier transistor were used as shown in Fig. 3-13. Initially the AGC adjust potentiometer was tested for shorted or open conditions, but checked out satisfactorily. Associated components of the two circuits were also tested without finding any defects in capacitors or resistors. An oscilloscope probe at the base of the gate transistor showed the arrival of the video signal from the video amplifier. Gating signals were observed at test point No. 2 of Fig. 3-13. Because no output signal was obtained from the AGC amplifier transistor, however, it was assumed this transistor was defective. The unit was removed and checked on a tester, which verified the fact that there was no signal transfer between input and output sections.

A new transistor restored full AGC function. The control was adjusted for a stable picture for the strongest station being received in the area.

12-40 PICTURE LOSS (SOUND UNAFFECTED)

In a color receiver there was no picture, though a raster was present. Sound for all stations was normal. Since sound and raster were present, it indicated that both the low-voltage and high-voltage supplies were producing power. Because of the dual detectors in a color receiver, as shown in Fig. 12-25, the loss of picture could be caused by a defective detector diode, or defective components and transistors from the video detector through the video amplifier stages to the picture tube. As shown, the separate sound IF detector was obtained directly from the output of

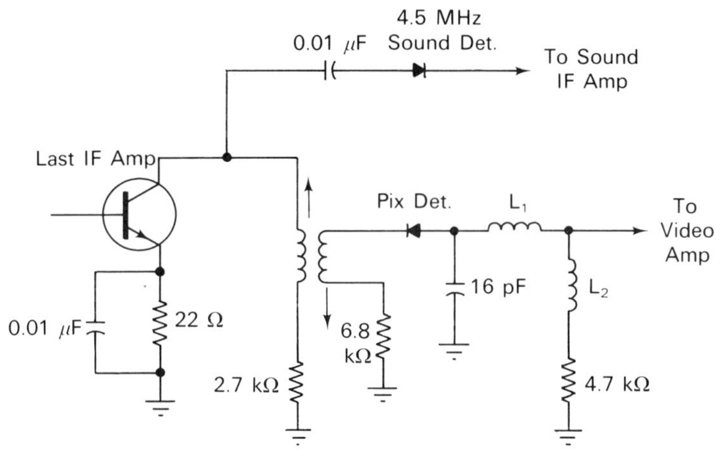

Figure 12-25 Detectors and components.

the last video IF amplifier, hence would not be disturbed by faults following the interstage IF transformer.

In the detector circuit two resistors isolate the diode from ground by over 10 KΩ, hence with the receiver shut off the forward-to-back ratio of the detector was tested while in the circuit. The diode showed a very low resistance in either polarity direction. (A shorted 16 pF capacitor could contribute to a resistance shunting effect across the diode, but in this instance the 6.8 kΩ resistor would still give a higher reading than a few ohms caused by a diode short.)

For this receiver the installation of a new picture-detector diode restored the picture to normal. If the diode circuitry had checked out normally, it would have been necessary to signal trace the following video amplifier stages to isolate the defective circuit. After that, the transistor (or tube) plus associated components would have to be checked to isolate the faulty unit.

12-41 WEAK SOUND AND PICTURE

In another receiver having circuits similar to those shown in Fig. 12-25, the picture and sound were present, but both were weak. Brilliancy and focus appeared to be all right, though the contrast control was ineffective in bringing up the picture strength. Because the sound was also weak, it was assumed that the trouble existed between the antenna and tuner system through the IF stages to the input of both detectors.

Instead of signal tracing initially from the tuner, a quick check was made by placing the scope at the output of the last video amplifier with a strong station tuned in. Full video signal information appeared here, indicating the faults were beyond this point. A check of the 0.01 μF coupling capacitor to the sound detector indicated this unit had changed value and was almost an open circuit. Sufficient capacitance remained to provide some coupling; hence sound was still present, though weak. A new capacitor restored full-volume sound, but the picture contrast was still below normal.

A scope test at the secondary of the interstage transformer indicate a weak signal, so the upper and lower tuning slugs of the transfer were adjusted to ascertain whether they were tuned properly. The original positions, however, were correct and the tuning of either core decreased the signal an additional amount. The diode and resistors were checked and found to be normal, but the 16 pF capacitor was defective and had a leakage resistance of approximately 3 kΩ. Thus, the leakage resistance was actually in shunt with the 4.7 kΩ output video load resistor, reducing the total load resistance to a value below that needed for maximum signal output. (Inductors L_1 and L_2, the peaking coils, have a negligible resistance.) Installation of a new 16 pF capacitor restored full picture strength.

This is an example of the dual troubles that occur on occasion as mentioned earlier and must always be suspected even when only a single symptom appears. (Loss of sound, for instance, could be caused by both a defective detector diode and an open series capacitor. Similarly, loss of picture could be caused by a defective detector diode or a component (or transistor) defect in a subsequent video amplifier stage.)

appendixes

A

Ohm's law factor

Because tests, measurements and troubleshooting procedures often involve use of basic electric-electronic values of current, voltage, power and resistance, some fundamental factors are reviewed in this section to serve as a ready reference when needed. Symbols, abbreviations, and terms now follow the International System of Units (designated as SI). The symbol for current is *I*, and the unit value (ampere) has the symbol A. Thus, current values of 0.05, 30 and 80 microamperes are expressed as 50 mA, 30 A, and 60 μA. The symbol for electromotive force is *E* and its unit value, the volt, has for its symbol V. Thus we have expressions such as 350 V, 30 μV, and 200 MV (megavolts).

Power has for its symbol *P*, while the unit value, the watt, is W, providing expressions such as 500 W, 30 mW (milliwatts) and 50 MW (megawatts). Resistance has for its symbol *R*, with the unit value, the ohm, given as Ω. Hence we have combinations such as 5Ω, 20 kΩ (kilohms) and 100 MΩ (megohms). (It must be remembered that *kilo* represents one thousand and *mega* one million, while *milli* designates *one-thousandth* and *micro* indicates *one millionth*) (see Appendix B).

The mathematical relationships of basic electric-electronic values are easily found by using the Ohm's law wheel shown in Fig. A-1. Thus, for solving the unknown values of R we can divide E by I, P by I^2, etc.

In alternating-current ratings, *reactance* (symbol X) refers to the opposition to current flow offered by capacitors or inductors. Because of opposing phase relationships between current and voltage for a reactance, no electric power is consumed by the reactance. However, when resistance is combined with reactance power is used. The opposition offered by a combination of resistance and reactance is known as *impedence* (symbol Z). The reactance and impedance factors are also related to Ohm's law. Thus, for ac we have expressions such as $Z = E/I$, $I = E/Z$, $X = E/I$, $E = IX$ (or IZ) and so forth.

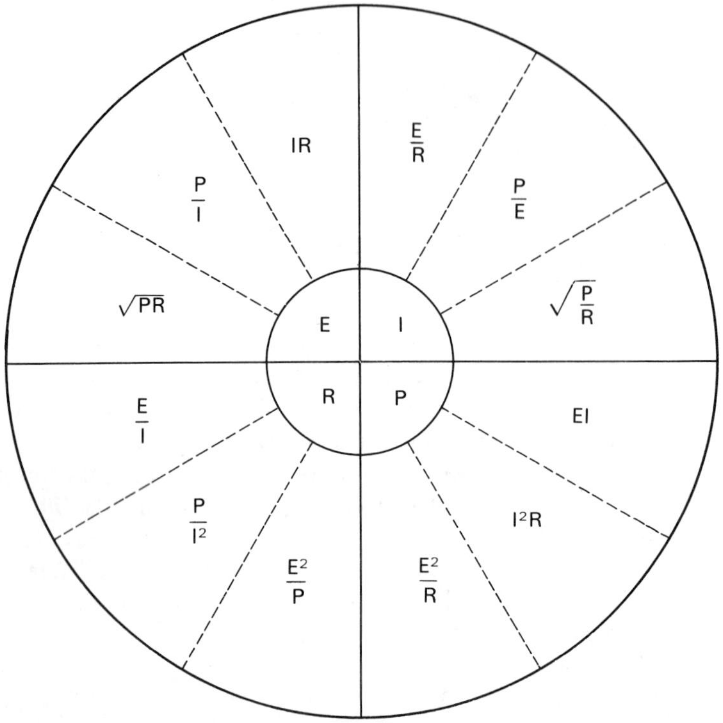

Figure A-1 Ohm's law wheel.

B

powers of ten factors

$$10 = 10^1$$
$$100 = 10^2$$
$$1{,}000 = 10^3$$
$$10{,}000 = 10^4$$
$$100{,}000 = 10^5$$
$$1{,}000{,}000 = 10^6$$
$$0.1 = 10^{-1}$$
$$0.01 = 10^{-2}$$
$$0.001 = 10^{-3}$$
$$\text{millionth} = 10^{-6}$$
$$\text{millionth of a millionth} = 10^{-12}$$
$$100^0 = 1$$
$$100^1 = 100$$
$$100^2 = 10{,}000$$
$$100^3 = 1{,}000{,}000$$

atto (a)		1×10^{-18}
femto (f)		1×10^{-15}
pico (p)	one-millionth of a millionth	1×10^{-12}

nano (n)	1/1000 of a millionth	1×10^{-9}
micro (μ)	one millionth	1×10^{-6}
milli (m)	one thousandth	1×10^{-3}
centi (c)	one hundredth	1×10^{-2}
deci (d)	one tenth	1×10^{-1}
deca (da)	ten	1×10^{1}
hecto (h)	one hundred	1×10^{2}
kilo (k)	one thousand	1×10^{3}
mega (M)	one million	1×10^{6}
giga (G)	one thousand million	1×10^{9}
tera (T)	one million million	1×10^{12}

C

public-entertainment broadcast allocations

AM RADIO-STATION ALLOCATIONS

550 kHz to 1,600 kHz
(Nominally 10 kHz bandwidth per station)
Preferred intermediate frequency (IF) 455 kHz.

FM RADIO-STATION ALLOCATIONS

88 MHz to 108 MHz
(200 kHz bandwidth per station)
Preferred intermediate frequency (IF) 10.7 MHz.

VHF TELEVISION STATION ALLOCATIONS

(6 MHz total bandwidth, including video and audio)
Preferred intermediate frequency (IF):
Picture carrier—45.75 MHz

Sound carrier—41.25 MHz
Pix and sound IF heterodyned to produce final sound IF of 4.5 MHz

Channel Number	*Frequency (MHz)*	*Video Carrier*	*Sound Carrier*
1	Not used		
2	54-60	55.25	59.75
3	60-66	61.25	65.75
4	66-72	67.25	71.75
5	76-82	77.25	81.75
6	82-88	83.25	87.75
	—FM Band (88 MHz to 108 MHz)—		
7	174-180	175.25	179.75
8	180-186	181.25	185.75
9	186-192	187.25	191.75
10	192-198	193.25	197.75
11	198-204	199.25	203.75
12	204-210	205.25	209.75
13	210-216	211.25	215.75

UHF Television Station Frequencies

Channel Number	*Frequency Range, MHz*	*Picture Carrier, MHz*	*Sound Carrier, MHz*
14	470-476	471.25	475.75
15	476-482	477.25	481.75
16	482-488	483.25	487.75
17	488-494	489.25	493.75
18	494-500	495.25	499.75
19	500-506	501.25	505.75
20	506-512	507.25	511.75
21	512-518	513.25	517.75
22	518-524	519.25	523.75
23	524-530	525.25	529.75
24	530-536	531.25	535.75
25	536-542	537.25	541.75
26	542-548	543.25	547.75
27	548-554	549.25	553.75
28	554-560	555.25	559.75

Channel Number	*Frequency Range, MHz*	*Picture Carrier, MHz*	*Sound Carrier, MHz*
29	560-566	561.25	565.75
30	566-572	567.25	571.75
31	572-578	573.25	577.75
32	578-584	579.25	583.75
33	584-590	585.25	589.75
34	590-596	591.25	595.75
35	596-602	597.25	601.75
36	602-608	603.25	607.75
37	608-614	609.25	613.75
38	614-620	615.25	619.75
39	620-626	621.25	625.75
40	626-632	627.25	631.75
41	632-638	633.25	637.75
42	638-644	639.25	643.75
43	644-650	645.25	649.75
44	650-656	651.25	655.75
45	656-662	657.25	661.75
46	662-668	663.25	667.75
47	668-674	669.25	673.75
48	674-680	675.25	679.75
49	680-686	681.25	685.75
50	686-692	687.25	691.75
51	692-698	693.25	697.75
52	698-704	699.25	703.75
53	704-710	705.25	709.75
54	710-716	711.25	715.75
55	716-722	717.25	721.75
56	722-728	723.25	727.75
57	728-734	729.25	733.75
58	734-740	735.25	739.75
59	740-746	741.25	745.75
60	746-752	747.25	751.75
61	752-758	753.25	757.75
62	758-764	759.25	763.75
63	764-770	765.25	769.75
64	770-776	771.25	775.75
65	776-782	777.25	781.75
66	782-788	783.25	787.75
67	788-794	789.25	793.75
68	794-800	795.25	799.75

Channel Number	*Frequency Range, MHz*	*Picture Carrier, MHz*	*Sound Carrier, MHz*
69	800-806	801.25	805.75
70	806-812	807.25	811.75
71	812-818	813.25	817.75
72	818-824	819.25	823.75
73	824-830	825.25	829.75
74	830-836	831.25	835.75
75	836-842	837.25	841.75
76	842-848	843.25	847.75
77	848-854	849.25	853.75
78	854-860	855.25	859.75
79	860-866	861.25	865.75
80	866-872	867.25	871.75
81	872-878	873.25	877.75
82	878-884	879.25	883.75
83	884-890	885.25	889.75

D

television signal data

In television transmission, various signals are combined to form a *composite* signal containing picture information and various pulses for synchronizing the beam sweep within the picture tube. The standard television composite signal is shown in Fig. 10-3. Horizontal blanking pulses blank out the screen during the retrace of the electron beam as it sweeps across the picture tube. Synchronizing pulses are mounted on the flat portion of the blanking pulse (termed the *pedestal*) to lock in the sweep circuits of the receiver. Picture (video) information occurs between blanking pulses. The picture tube beam sweeps across the picture tube and progressively downward forming what is termed a *field*. Alternate fields interlace with preceding ones, and two fields form a *frame*. Equalizing pulses maintain horizontal synchronization during vertical blanking (when the beam retraces from the bottom of the screen to the top). Vertical sync pulses consist of six blocks so as to maintain horizontal synchronization during the vertical beam control.

One frame = 33,334 μs (microseconds)
One field = 16,667 μs

One horizontal sweep cycle (start of one horizontal line trace to start of next) = 63.5 μs
Horizontal blanking interval = 10.16 μs to 11.4 μs
Horizontal trace (without blanking time) = 53.34 μs
Horizontal sync pulse duration = 5.08 μs to 5.68 μs
Vertical sync pulse interval (total of six vertical blocks) = 190.5 μs
Vertical blanking interval = 833 μs to 1300 μs for each field
Vertical scan frequency (monochrome) = 60 Hz
Vertical scan frequency (color) = 59.94 Hz
Horizontal scan frequency (monochrome) = 15,750 Hz
Horizontal scan frequency (color) = 15,734.264 Hz
Total frequency span for an individual station (monochrome or color) = 6 MHz
Picture carrier is nominally 1.25 MHz above the lower end of the channel (monochrome or color)
Vestigial sideband transmission is used for both monochrome and color
Aspect ratio of monochrome or color picture is 4:3
Scan lines per frame (mono or color) = 525 (interlaced)
Scan lines per field (mono or color) = 262.5
The frequency-modulated sound carrier is 4.5 MHz above the picture-carrier frequency (mono or color), with maximum deviation 25 kHz each side of center frequency
The effective radiated power of the audio may range from 50 to 70 percent of the peak power of the picture-signal carrier

E

color television factors and terms

The color signal video corresponds to a brightness component transmitted as an amplitude-modulated (AM) video carrier, plus a pair of color signals transmitted as AM sidebands procured from a phase-modulated carrier. Hence, two subcarriers 90° apart were involved, having a common frequency related to the color-picture carrier frequency of 3.579,545 MHz.

The modulated color carriers are suppressed at the transmitter, and only sideband signals are telecast. An independent carrier oscillator is used at the receiver and synchronized by a color burst signal on the horizontal pedestal. A minimum of eight cycles constitute the burst signal, the frequency of which is identical to the one used at the transmitter: 3.579,545 MHz. The burst signal is omitted following equalizing pulses and is also absent during the vertical pulse interval (or during monochrome transmission).

The brightness, or luminance, portion of the video signal is also referred to as the Y signal.

Hue is the term which identifies a color (such as blue hue, green hue, etc.)

Brightness refers to the intensity of a color.

Saturation refers to the degree to which a color is diluted with white. The less white a color contains, the higher its saturation.

The Y signal consists of predetermined levels of each of the primary colors of red, blue and green, to correspond to the proportions to which the average eye responds. The Y signal thus contains: 0.59 green, 0.30 red, and 0.11 blue.

The I signal consists of combinations of B-Y and R-Y signals, and the proportions are: −0.27 blue − Y; and 0.74 red − Y.

The Q signal (also B-Y and R-Y) consists of: 0.41 blue − Y, and 0.48 red − Y.

F

trigonometric ratios

Degrees	$\frac{(opp)}{(hyp)}$ Sine	$\frac{(adj)}{(hyp)}$ Cosine	$\frac{(opp)}{(adj)}$ Tangent	Degrees	$\frac{(opp)}{(hyp)}$ Sine	$\frac{(adj)}{(hyp)}$ Cosine	$\frac{(opp)}{(adj)}$ Tangent
0	0.0000	1.0000	0.0000	15	.2588	.9659	.2679
1	.0175	.9998	.0175	16	.2756	.9613	.2867
2	.0349	.9994	.0349	17	.2924	.9563	.3057
3	.0523	.9986	.0524	18	.3090	.9511	.3249
4	.0698	.9976	.0699	19	.3256	.9455	.3443
5	.0872	.9962	.0875	20	.3420	.9397	.3640
6	.1045	.9945	.1051	21	.3584	.9336	.3839
7	.1219	.9925	.1228	22	.3746	.9272	.4040
8	.1392	.9903	.1405	23	.3907	.9205	.4245
9	.1564	.9877	.1584	24	.4067	.9135	.4452
10	.1736	.9848	.1763	25	.4226	.9063	.4663
11	.1908	.9816	.1944	26	.4384	.8988	.4877
12	.2079	.9781	.2126	27	.4540	.8910	.5095
13	.2250	.9744	.2309	28	.4695	.8829	.5317
14	.2419	.9703	.2493	29	.4848	.8746	.5543

Degrees	(opp)/(hyp) Sine	(adj)/(hyp) Cosine	(opp)/(adj) Tangent	Degrees	(opp)/(hyp) Sine	(adj)/(hyp) Cosine	(opp)/(adj) Tangent
30	.5000	.8660	.5774	60	0.8660	0.5000	1.7321
31	.5150	.8572	.6009	61	.8746	.4848	1.8040
32	.5299	.8480	.6249	62	.8829	.4695	1.8807
33	.5446	.8387	.6494	63	.8910	.4540	1.9626
34	.5592	.8290	.6745	64	.8988	.4384	2.0503
35	.5736	.8192	.7002	65	.9063	.4226	2.1445
36	.5878	.8090	.7265	66	.9135	.4067	2.2460
37	.6018	.7986	.7536	67	.9205	.3907	2.3559
38	.6157	.7880	.7813	68	.9272	.3746	2.4751
39	.6293	.7771	.8098	69	.9336	.3584	2.6051
40	.6428	.7660	.8391	70	.9397	.3420	2.7475
41	.6561	.7547	.8693	71	.9455	.3256	2.9042
42	.6691	.7431	.9004	72	.9511	.3090	3.0777
43	.6820	.7314	.9325	73	.9563	.2924	3.2709
44	.6947	.7193	.9657	74	.9613	.2756	3.4874
45	.7071	.7071	1.0000	75	.9659	.2588	3.7321
46	.7193	.6947	1.0355	76	.9703	.2419	4.0108
47	.7314	.6820	1.0724	77	.9744	.2250	4.3315
48	.7431	.6691	1.1106	78	.9781	.2079	4.7046
49	.7547	.6561	1.1504	79	.9816	.1908	5.1446
50	.7660	.6428	1.1918	80	.9848	.1736	5.6713
51	.7771	.6293	1.2349	81	.9877	.1564	6.3138
52	.7880	.6157	1.2799	82	.9903	.1392	7.1154
53	.7986	.6018	1.3270	83	.9925	.1219	8.1443
54	.8090	.5878	1.3764	84	.9945	.1045	9.5144
55	.8192	.5736	1.4281	85	.9962	.0872	11.4301
56	.8290	.5592	1.4826	86	.9976	.0698	14.3007
57	.8387	.5446	1.5399	87	.9986	.0523	19.0811
58	.8480	.5299	1.6003	88	.9994	.0349	28.6363
59	.8572	.5150	1.6643	89	.9998	.0175	57.2900
60	.8660	.5000	1.7321	90	1.000	.0000	......

G

resistor color coding

Molded-composition resistors have leads at each end as shown in Fig. A-2 and are available in various wattage ratings such as $\frac{1}{4}$ W, $\frac{1}{2}$ W, and 1 W, etc. The ohmic value of each resistor is indicated by standard color-coded bands grouped at one end of the resistor are identified starting at that end as shown in Fig. A-2. Usually four bands of color are employed for the carbon composition resistors, and five bands are used for the film types. In either case, the last color denotes the *tolerance* that must be applied to the value obtained from reading the initial bands. Thus, if a rated 100-Ω resistor had a tolerance of 10 percent, its actual ohmic value could range from 90 to 110 Ω.

The listing that follows shows the color-code values and tolerances that apply to the carbon and film resistors shown in Fig. A-2. The abbreviation GMV indicates *guaranteed minimum value.* When a value is marked "alternate" (*alt.*), it indicates a coding that may have been used occasionally in the past but is generally referred to in modern components by the "preferred" (*pref.*) coding designation.

The standard listing for the resistor color coding follows and a typical coding example is shown in Fig. A-2.

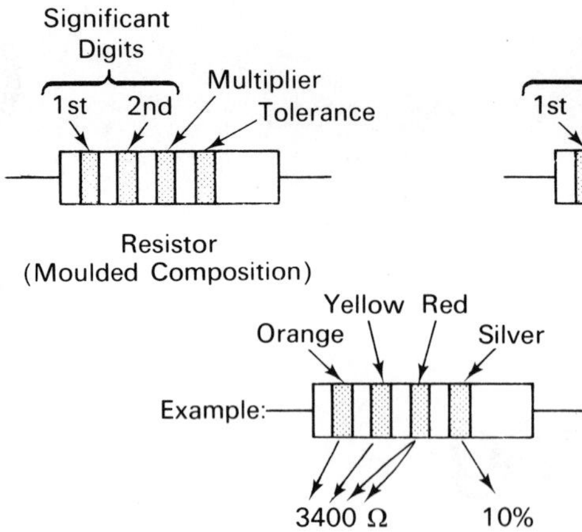

Figure A-2 Resistor color coding.

Resistance in Ohms (Ω)

Color	*Digit*	*Multiplier*	*Carbon ± Tolerance*	*Film-type Tolerance*
Black	0	1	20%	0
Brown	1	10	1%	1%
Red	2	100	2%	2%
Orange	3	1000	3%	
Yellow	4	10,000	GMV	
Green	5	100,000	5% *(alt.)*	0.5%
Blue	6	1,000,000	6%	0.25%
Violet	7	10,000,000	12.5%	0.1%
Gray	8	0.01 *(alt.)*	30%	0.05%
White	9	0.1 *(alt.)*	10% *(alt.)*	
Silver		0.01 *(pref.)*	10% *(pref.)*	10%
Gold		0.1 *(pref.)*	5% *(pref.)*	5%
No color			20%	

H

capacitor color codings

The color codings for various ceramic-type capacitors are shown in Figs. A-3 and A-4. As shown in Fig. A-3, the tubular types may have *axial leads* (i.e., leads from the ends) or *radial leads* (i.e., leads connected at right angles to the capacitor length). The values that are read from the digit color are in picofarads (pF). A *temperature-coefficient band* is also present in addition to the tolerance color band and the digit bands.

The coefficient of the ceramic capacitors is given in parts per million per degree centigrade (ppm/°C). A letter N preceding this value denotes *negative-temperature coefficient;* NPO designates a *negative-positive-zero coefficient.* Thus, a designation such as N220 indicates a capacitance decrease with a rise in temperature of 220 parts per million per °C and shows by how much the value changes during the warm-up time of the device in which this capacitor is used. The NPO types are stable units with negligible temperature effect on capacitance values.

As shown in Fig. A-3, five identification markings are used for these capacitors. For the axial-lead type, the identification starts at the end at which the color bands are grouped. The first band represents the temperature coefficient marking, and the next two bands represent the significant digits making up the value.

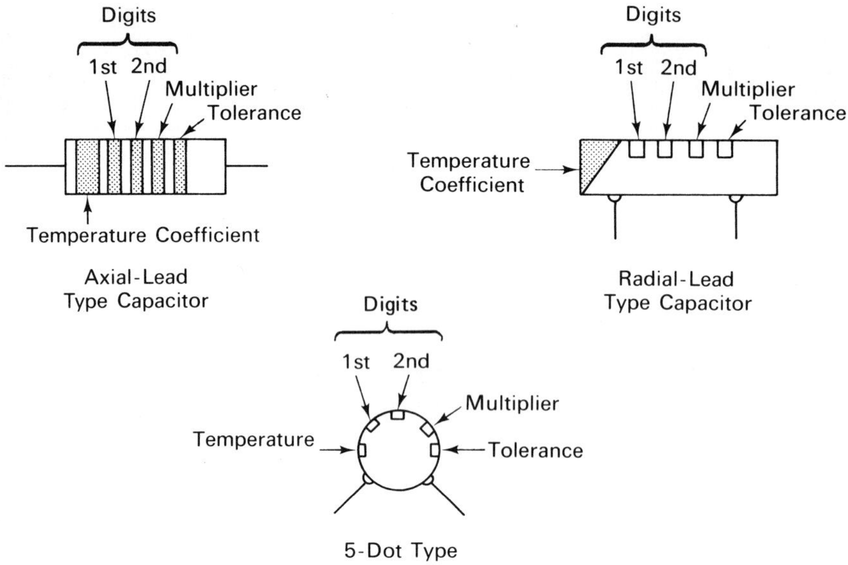

Figure A-3 Ceramic capacitor color coding.

Often, the axial-type ceramic capacitors have a wider first band for identification purposes. The radial types have an initial band of greater color area as shown. For the five-dot disc type shown in Fig. A-3, the lower left color dot gives the temperature coefficient; the other dots in clockwise sequence have the same coding as do the axial or radial types. The button-silver mica, the button ceramic, and the feed-through ceramic types are shown in Fig. A-4.

The extended-range temperature coefficient ceramic capacitor is shown in Fig. A-5 with a molded mica type. For the extended-range type, the first color sector indicates the temperature coefficient as is the case with the five-dot types, but the second color segment represents the temperature-coefficient multiplier. The following initial listing applies to the five-dot and extended-range types, including the disc types. For those disc types that have only three color dots, the temperature coefficient and tolerance values are not given. In these, the first two dots (clockwise) are the significant digits and the last dot is the multiplier, as shown for the button ceramic in Fig. A-4.

The second listing given applies to the flat, rectangular-shaped molded mica type capacitor shown in Fig. A-5. An arrow, or an arrowhead, is imprinted on the capacitor face to indicate the direction of color-coding sequence. The lower left-hand color dot shows the type or classification of the particular capacitor according to the manufacturer's

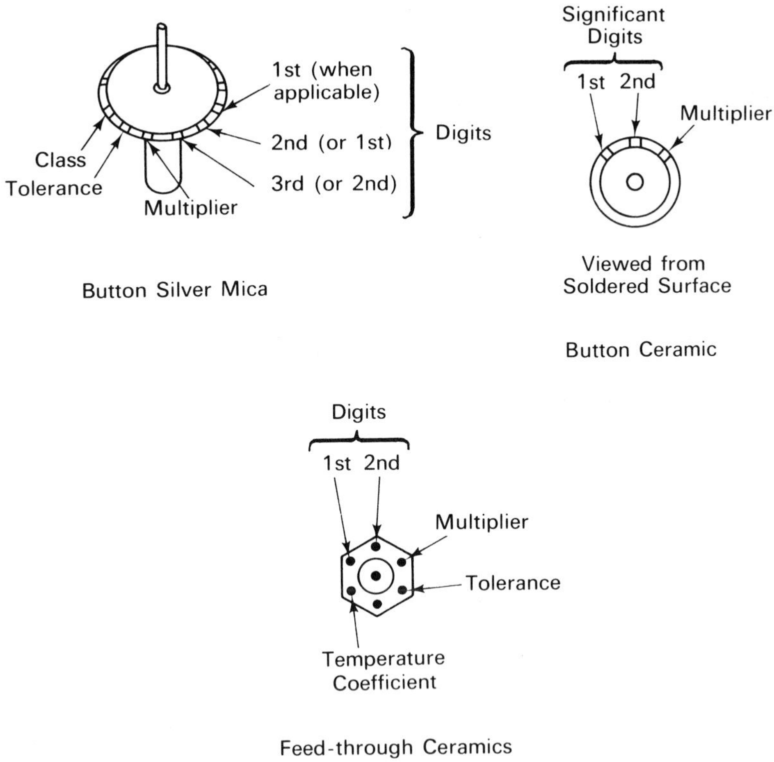

Figure A-4 Button and feed-through ceramic coding.

specifications, regarding temperature coefficient, Q factor, and other related characteristics.

Ceramic Types (Capacitance in Picofarads)

Color	*Digit*	*Multiplier*	*10 pF or less*	*Over 10 pF*	*5-dot Temp. Coeff. TC*	*Extended Range Significant Digits*	*Extended Range Multiplier*
Black	0	1	2.0 pF	20%	NP0	0.0	−1
Brown	1	10	0.1 pF	1%	N033		−10
Red	2	100		3%	N075	1.0	−100
Orange	3	1000		3%	N150	1.5	−1000
Yellow	4	10,000			N220	2.2	−10,000
Green	5		0.5 pF	5%	N330	3.3	+1

Color	Digit	Multiplier	10 pF or less	Over 10 pF	5-dot Temp. Coeff. TC	Extended Range Significant Digits	Extended Range Multiplier
Blue	6				N470	4.7	+10
Violet	7				N750	7.5	+100
Gray	8	0.01 (alt.)	0.25 pF		*	†	+1000
White	9	0.1 (alt.)	1.0 pF	10%	**		+10,000
Silver		0.01 (pref.)					
Gold		0.1 (pref.)					

* General-purpose types with a TC ranging from P150 to N1500.
** Coupling, decoupling and general bypass with a TC ranging from P100 to N750.
† If the first band (TC) is black, the range is N1000 to N5000.

Mica Types (Capacitance in Picofarads)

Color	Digit	Multiplier	Tolerance	Type Classification
Black	0	1	20%(±)	A
Brown	1	10	1%	B
Red	2	100	2%	C
Orange	3	1000	3%	D
Yellow	4	10,000		E
Green	5		5%	
Blue	6			
Violet	7			
Gray	8			
White	9		10%	
Silver		0.01		
Gold		0.1		

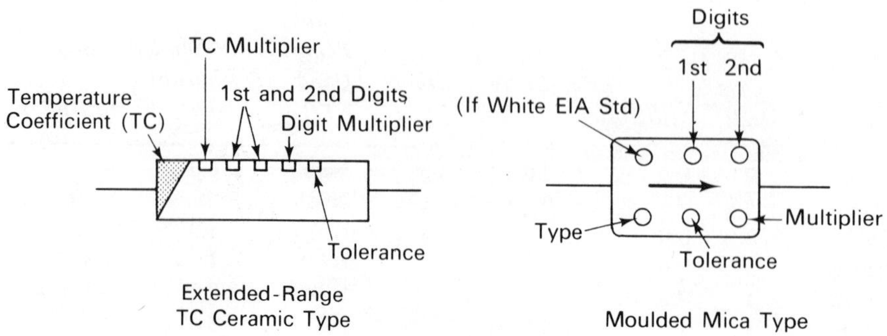

Figure A-5 Extended range and mica capacitor codings.

I

transformer color codings

Color codings for transformers are shown in Fig. A-6 and apply to power transformers, audio transformers and RF and IF transformers in re-

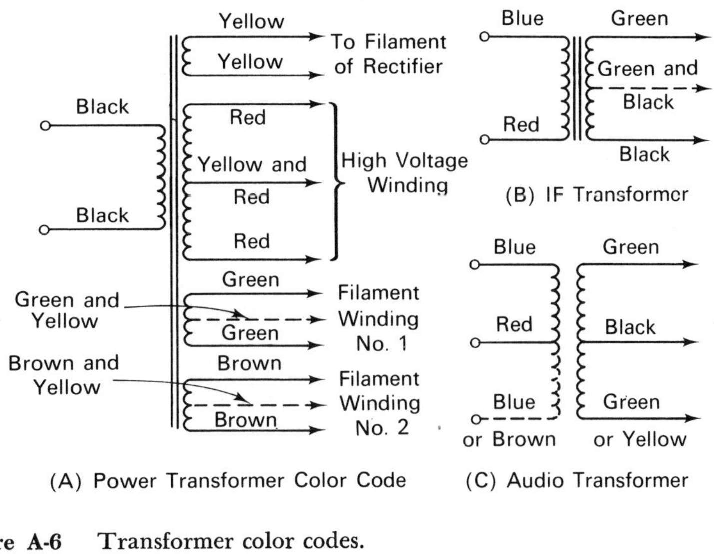

Figure A-6 Transformer color codes.

ceivers. Thus, for those manufacturers who follow established standards, various primary and secondary leads are colored to conform to the type of terminal represented by the wire leads from the transformer. The audio transformer shown in Fig. A-6C may be an interstage type connected from the output of small-signal transistors (or tubes) to the input circuitry of output transistors or tubes. These types could also have been output transformers connecting the output from amplifiers through the transformer to the voice coils of loudspeakers or earphones.

index

H

I

L

M

N

O

U

V

W

Y